OPTICAL WAVEGUIDE THEORY BY THE FINITE ELEMENT METHOD

Advances in Optoelectronics (ADOP)

ADOP   Advances in Optoelectronics

OPTICAL WAVEGUIDE THEORY BY THE FINITE ELEMENT METHOD

Masanori KOSHIBA
Hokkaido University, Sapporo

Springer-Science+Business Media, B.V.

Library of Congress Cataloging in Publication Data

CIP DATA APPEARS ON SEPARATE CARD

ISBN 978-0-7923-2080-7   ISBN 978-94-011-1634-3 (eBook)
DOI 10.1007/978-94-011-1634-3

PREFACE

Recent advances in the field of guided-wave optics, such as fiber optics and integrated optics, have included the introduction of arbitrarily-shaped optical waveguides which, in many cases, also happened to be arbitrarily inhomogeneous, dissipative, anisotropic, and/or nonlinear. Most of such cases of waveguide arbitrariness do not lend themselves to analytical solutions; hence, computational tools for modeling and simulation are essential for successful design, optimization, and realization of the optical waveguides. For this purpose, various numerical techniques have been developed. In particular, the finite element method (FEM) is a powerful and efficient tool for the most general (i.e., arbitrarily-shaped, inhomogeneous, dissipative, anisotropic, and nonlinear) optical waveguide problem. Its use in industry and research is extensive, and indeed it could be said that without it many optical waveguide problems would be incapable of solution.

This book is intended for students, engineers, designers, and technical managers interested in a detailed description of the FEM for optical waveguide analysis.

Starting from a brief review of electromagnetic theory, the first chapter provides the concepts of the FEM and its fundamentals. In addition to conventional elements, i.e., line elements, triangular elements, tetrahedral elements, ring elements, and triangular ring elements which are utilized for one-dimensional, two-dimensional, three-dimensional, axisymmetric two-dimensional, and axisymmetric three-dimensional problems, respectively, special-purpose elements, such as isoparametric elements, edge elements, infinite elements, and boundary elements, are also introduced. Chapters 2 to 5 establish the theory of planar optical waveguides, optical channel waveguides, optical fibers, and polarization-maintaining optical fibers. Chapters 6 and 7 describe the diffraction characteristics of optical gratings and the scattering characteristics of optical waveguide discontinuities, respectively. Chapter 8 is devoted to a description of nonlinear optical waveguides, placing emphasis on the second-harmonic generation and intensity-dependent refractive-index phenomena. Chapter 9 deals with optical solitons which may find future practical applications in large-capacity, long-distance transmission systems. Finally, Chapter 10 presents a simple description of the quantum well, superlattice, and resonant tunneling structures which have been attracting much attention in recent years because of their electronic and optoelectronic properties.

All the illustrative numerical examples which will be presented in this book have been calculated by the author and his co-workers, whose papers, written in English, are cited in each chapter of this volume.

The author would like to express his gratitude to Professor T. Okoshi of the University of Tokyo, who gave him the opportunity and encouragement to write this book.

The author also wishes to thank Associate Professor K. Hayata of Hokkaido University and Associate Professor K. Hirayama of Kitami Institute of Technology for their helpful discussions and suggestions.

His thanks are also extended to those students who worked in his laboratory at Hokkaido University for their Ph.D., M.S., or B.S. theses.

July, 1992

Masanori KOSHIBA

Contents

Chapter 1

FINITE ELEMENT METHOD

1.1 Introduction

In the finite element method (FEM),[1)-5)] instead of differential equations (governing equations) for the system under consideration, corresponding functionals (variational expressions) to which a variational principle is applied are set up, where the region of interest is divided into the so-called elements; an equivalent discretized model for each element is constructed; and then all the element contributions to the system are assembled. In other words, the FEM can be considered a subclass of the Rayleigh-Ritz method, in which piecewise defined polynomial functions are used for trial functions and infinite degrees of freedom of the system are discretized or replaced by a finite number of unknown parameters. In classic analytical procedures without subdivision processes, the system is modeled using analytical functions defined over the whole region of interest, and therefore these procedures generally are applicable only to simple geometries and materials. Of the various forms of discretization possible, one of the simplest is the finite difference method (FDM). And its traditional versions use a regular grid; that is, a rectangular grid with nodes at the intersections of orthogonal straight lines. However, a regular grid is not suitable for curved boundaries or interfaces, because they intersect gridlines obliquely at points other than the nodes. Moreover, a regular grid is not suitable for problems with very steep variations of fields. The FEM is somewhat similar to the FDM. In the FEM, the field region is subdivided into elements; that is, into subregions. Elements can have various shapes, such as triangles and rectangles, allowing the use of an irregular grid. Therefore, the FEM is suitable for problems with very steep variations of fields. Furthermore, this approach can be easily adapted to inhomogeneous and anisotropic problems, and it is possible to systematically increase the accuracy of the solutions obtained, as necessary. Furthermore, the finite element scheme can be established

1

not only by the variational method but by the Galerkin method, which is a
weighted residual method. Therefore, the FEM may be applicable to problems where a variational principle does not exist or cannot be identified.

In this chapter the concept of the FEM and its fundamentals are dis-
cussed. This includes a listing of the relevant forms of Maxwell's equations
for optical waveguide analysis.

1.2   Maxwell's Equations

Maxwell's equations for time-dependent fields are

$$\nabla \times \boldsymbol{E} \; = \; -\partial \boldsymbol{B}/\partial t \tag{1.1}$$

$$\nabla \times \boldsymbol{H} \; = \; \partial \boldsymbol{D}/\partial t + \boldsymbol{J} \tag{1.2}$$

$$\nabla \cdot \boldsymbol{D} \; = \; \rho \tag{1.3}$$

$$\nabla \cdot \boldsymbol{B} \; = \; 0 \tag{1.4}$$

where $\boldsymbol{E}$ and $\boldsymbol{H}$ are the electric and magnetic fields, respectively, $\boldsymbol{D}$ and
$\boldsymbol{B}$ are the electric flux and magnetic flux densities, respectively, $\boldsymbol{J}$ is the
electric current density, and ρ is the electric charge density. These symbolic
equations are frequently displayed in component form by means of a matrix
representation, with fields described by the column matrices

$$\boldsymbol{E} = \begin{bmatrix} E_x \\ E_y \\ E_z \end{bmatrix} , \quad \text{etc.} \tag{1.5}$$

and the curl and divergence operators described by the matrix-differential
operators

$$\nabla \times \; = \; \begin{bmatrix} 0 & -\partial/\partial z & \partial/\partial y \\ \partial/\partial z & 0 & -\partial/\partial x \\ -\partial/\partial y & \partial/\partial x & 0 \end{bmatrix} \tag{1.6}$$

$$\nabla \cdot \; = \; \begin{bmatrix} \partial/\partial x & \partial/\partial y & \partial/\partial z \end{bmatrix} \tag{1.7}$$

in Cartesian coordinates. These operators have the matrix forms

$$\nabla \times \; = \; \begin{bmatrix} 0 & -\dfrac{\partial}{\partial z} & \dfrac{1}{r}\dfrac{\partial}{\partial \theta} \\[2ex] \dfrac{\partial}{\partial z} & 0 & -\dfrac{\partial}{\partial r} \\[2ex] -\dfrac{1}{r}\dfrac{\partial}{\partial \theta} & \dfrac{\partial}{\partial r}+\dfrac{1}{r} & 0 \end{bmatrix} \tag{1.8}$$

$$\nabla \cdot \; = \; \begin{bmatrix} \dfrac{\partial}{\partial r}+\dfrac{1}{r} & \dfrac{1}{r}\dfrac{\partial}{\partial \theta} & \dfrac{\partial}{\partial z} \end{bmatrix} \tag{1.9}$$

in cylindrical coordinates.

The Poynting vector $\boldsymbol{S}$ representing the energy flow is defined as

$$\boldsymbol{S} = \boldsymbol{E} \times \boldsymbol{H} \, . \tag{1.10}$$

Assuming an isotropic medium with a scalar permittivity ε and a scalar permeability μ, we have the constitutive relations

$$\boldsymbol{D} = \varepsilon \boldsymbol{E} = \varepsilon_0 \varepsilon_r \boldsymbol{E} \tag{1.11}$$

$$\boldsymbol{B} = \mu \boldsymbol{H} = \mu_0 \mu_r \boldsymbol{H} \tag{1.12}$$

where ε_0 and μ_0 are the permittivity and permeability of free space, respectively, ε_r is the relative permittivity, and μ_r is the relative permeability. The refractive index n is defined as

$$n = \sqrt{\varepsilon_r \mu_r} \, . \tag{1.13}$$

In optical waveguides we usually have a constant permeability $\mu = \mu_0$, which implies

$$n = \sqrt{\varepsilon_r} \, . \tag{1.14}$$

For an anisotropic medium with a tensor permittivity $[\varepsilon]$ the constitutive relation, Eq. (1.11), is replaced by

$$\boldsymbol{D} = [\varepsilon]\boldsymbol{E} = \varepsilon_0 [\varepsilon_r]\boldsymbol{E} \tag{1.15}$$

where $[\varepsilon_r]$ is the relative permittivity tensor.

In a nonlinear case polarizations induced in the medium can be expressed as

$$P_i = \varepsilon_0 (\chi_{ij} E_j + \chi_{ijk} E_j E_k + \chi_{ijkl} E_j E_k E_l + \cdots) \tag{1.16}$$

where P_i is the ith component of polarization, E_i is the ith component of electric field, χ_{ij} is the linear susceptibility, χ_{ijk} and χ_{ijkl} are the second-order and third-order nonlinear optical susceptibilities, respectively, and summations over repeated indices are assumed. Since no physical significance can be attached to the interchange of j and k in the second-order nonlinear optical susceptibilities, we can replace the subscripts jk and kj by the contracted indices:

$$xx = 1 \, , \quad yy = 2 \, , \quad zz = 3 \, ,$$
$$yz = zy = 4 \, , \quad zx = xz = 5 \, , \quad xy = yx = 6 \, .$$

The resulting components form a 3×6 matrix called the second-order nonlinear optical tensor $[d]$ that operates on the E^2 column matrix to yield the second-order nonlinear polarization $\boldsymbol{P}_{\mathrm{NL}}$ according to

$$
\boldsymbol{P}_{\mathrm{NL}} = \begin{bmatrix} P_x \\ P_y \\ P_z \end{bmatrix} = \varepsilon_0 \begin{bmatrix} d_{11} & d_{12} & d_{13} & d_{14} & d_{15} & d_{16} \\ d_{21} & d_{22} & d_{23} & d_{24} & d_{25} & d_{26} \\ d_{31} & d_{32} & d_{33} & d_{34} & d_{35} & d_{36} \end{bmatrix} \begin{bmatrix} E_x^2 \\ E_y^2 \\ E_z^2 \\ 2E_yE_z \\ 2E_zE_x \\ 2E_xE_y \end{bmatrix}
$$

(1.17)

where the nonlinear optical coefficients d_{11} to d_{36} are given by

$$
\begin{aligned}
&d_{11} = \chi_{xxx}\,, \quad d_{12} = \chi_{xyy}\,, \quad d_{13} = \chi_{xzz}\,, \\
&d_{14} = (\chi_{xyz} + \chi_{xzy})/2\,, \\
&d_{15} = (\chi_{xzx} + \chi_{xxz})/2\,, \\
&d_{16} = (\chi_{xxy} + \chi_{xyx})/2\,, \quad \text{etc.}
\end{aligned}
$$

(1.18)

Maxwell's equations are subject to boundary conditions at the interface where abrupt changes of the material constant occur. In the absence of interface charges and interface currents, we have the conditions

$$
\begin{aligned}
\boldsymbol{n} \times (\boldsymbol{E}_1 - \boldsymbol{E}_2) &= 0 & (1.19) \\
\boldsymbol{n} \times (\boldsymbol{H}_1 - \boldsymbol{H}_2) &= 0 & (1.20) \\
\boldsymbol{n} \cdot (\boldsymbol{D}_1 - \boldsymbol{D}_2) &= 0 & (1.21) \\
\boldsymbol{n} \cdot (\boldsymbol{B}_1 - \boldsymbol{B}_2) &= 0 & (1.22)
\end{aligned}
$$

where $\boldsymbol{n}$ is the unit vector normal to the interface directed from medium 1 into medium 2. In optical waveguides the permeability is assumed to be scalar and constant and so all components of magnetic field $\boldsymbol{H}$ are continuous everywhere.

Time-harmonic fields may be written in the form

$$
\boldsymbol{E}(\boldsymbol{r}, t) = [\boldsymbol{E}(\boldsymbol{r}) \exp(j\omega t) + \text{c.c.}]/2\,, \quad \text{etc.}
$$

(1.23)

where $\boldsymbol{E}(\boldsymbol{r})$ is a complex amplitude, $\boldsymbol{r}$ is the radius vector, ω is the angular frequency, and c.c. denotes a complex conjugate. Maxwell's equations for source free, time-harmonic fields in optical waveguides are then reduced to

$$
\begin{aligned}
\nabla \times \boldsymbol{E} &= -j\omega\mu_0\boldsymbol{H} & (1.24) \\
\nabla \times \boldsymbol{H} &= j\omega[\varepsilon]\boldsymbol{E} = j\omega\varepsilon_0[\varepsilon_r]\boldsymbol{E} & (1.25) \\
\nabla \cdot \boldsymbol{D} &= \nabla \cdot ([\varepsilon]\boldsymbol{E}) = \varepsilon_0\nabla \cdot ([\varepsilon_r]\boldsymbol{E}) = 0 & (1.26) \\
\nabla \cdot \boldsymbol{H} &= 0\,. & (1.27)
\end{aligned}
$$

The complex Poynting vector is defined as

$$S = \frac{1}{2}\boldsymbol{E} \times \boldsymbol{H}^* \tag{1.23}$$

where * indicates a complex conjugate.

1.3 Outline of Finite Element Calculation

To clarify ideas of the FEM, consider a particular problem governed by the Helmholtz equation.

The specific governing equation is now written for a domain Ω shown in Fig. 1.1 as

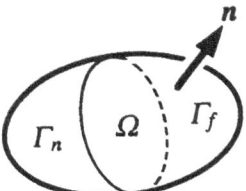

Figure 1.1: Three-dimensional region Ω surrounded by boundaries Γ_f and Γ_n.

$$\nabla^2\phi + k^2\phi = 0 \quad \text{in } \Omega \tag{1.29}$$

where ϕ is the electric or magnetic field component and the quantity k^2 is a constant related to frequency. The Laplacian operator is given by

$$\nabla^2 = \partial^2/\partial x^2 + \partial^2/\partial y^2 + \partial^2/\partial z^2 \tag{1.30}$$

in Cartesian coordinates. This operator has the form

$$\nabla^2 = \frac{1}{r}\frac{\partial}{\partial r}\left(r\frac{\partial}{\partial r}\right) + \frac{1}{r^2}\frac{\partial^2}{\partial\theta^2} + \frac{\partial^2}{\partial z^2} \tag{1.31}$$

in cylindrical coordinates.

Now, we assume that the boundary Γ of the region Ω consists partly of Γ_f on which the value of ϕ is given as $\hat{\phi}$ and partly of Γ_n on which the value of $\partial\phi/\partial n \equiv \psi$ is given as $\hat{\psi}$; namely,

$$\phi = \hat{\phi} \quad \text{on } \Gamma_f \tag{1.32}$$

$$\partial\phi/\partial n = \boldsymbol{n}\cdot\nabla\phi = \hat{\psi} \quad \text{on } \Gamma_n \tag{1.33}$$

where $\boldsymbol{n}$ is the outward unit normal vector. The gradient operator is described by the matrix-differential operator

$$\nabla = \begin{bmatrix} \partial/\partial x \\ \partial/\partial y \\ \partial/\partial z \end{bmatrix} \tag{1.34}$$

in Cartesian coordinates. This operator has the matrix form

$$\nabla = \begin{bmatrix} \dfrac{\partial}{\partial r} \\[2mm] \dfrac{1}{r}\dfrac{\partial}{\partial \theta} \\[2mm] \dfrac{\partial}{\partial z} \end{bmatrix} \tag{1.35}$$

in cylindrical coordinates.

The boundary conditions, Eqs. (1.32) and (1.33), are called Dirichlet and Neumann boundary conditions, respectively. Considering these boundary conditions, the functional for Eq. (1.29) is given by

$$F = \frac{1}{2}\iiint_{\Omega}[(\nabla\phi)^2 - k^2\phi^2]\, d\Omega - \iint_{\Gamma_n}\phi\hat{\psi}\, d\Gamma\ . \tag{1.36}$$

The first and second terms in the right-hand side of Eq. (1.36) denote the integrals over the region Ω and along the boundary Γ_n, respectively. The Euler equation derived by the stationary requirement

$$\delta F = 0 \tag{1.37}$$

coincides with the governing equation, namely, Eq. (1.29). The boundary condition in Eq. (1.33) is called the natural boundary condition, because this condition is automatically satisfied in the variational procedure. On the other hand, the boundary condition in Eq. (1.32) should be imposed on trial functions; therefore, this condition is called the forced boundary condition.

Dividing the region Ω into a number of elements e and considering the functional F_e

$$F_e = \frac{1}{2}\iiint_{e}[(\nabla\phi)^2 - k^2\phi^2]\, d\Omega - \iint_{\Gamma_e}\phi\psi\, d\Gamma \tag{1.38}$$

for each element, the functional for the whole region is given by

$$F = \sum_e F_e\ . \tag{1.39}$$

The first and second terms in the right-hand side of Eq. (1.38) denote the integrals over each element e and along the element boundary Γ_e, respectively, and the summation $\sum_e$ extends over all different elements.

Arranging n nodes in each element, ϕ can be approximated as

$$\phi = \sum_{i=1}^{n} N_i \phi_i \qquad (1.40)$$

where ϕ_i is the ith nodal parameter of the element e and N_i is the interpolation or shape function.

When the functional value contains first-order derivatives, to guarantee the convergence of solutions, the shape function N_i should satisfy the following two conditions:

(1) The variable ϕ and its derivatives must include the constant terms.

(2) The variable ϕ must be continuous at the interface between two adjacent elements.

These are called the completeness and compatibility conditions, respectively. The completeness condition is simple to satisfy if complete polynomial expressions are used in each element. First-order or linear elements are the most fundamental ones, and use first-order polynomials. Higher-order elements, on the other hand, use higher-order polynomials. The number of nodes within each element, n, coincides with the number of terms in a complete polynomial expansion, and nodes are arranged to satisfy the compatibility condition.

We can express Eq. (1.40) in matrix form:

$$\phi = \{N\}^{\mathrm{T}} \{\phi\}_e \qquad (1.41)$$

where the components of the $\{\phi\}_e$ and $\{N\}$ vectors are ϕ_i and N_i, respectively, and T, $\{\cdot\}$, and $\{\cdot\}^{\mathrm{T}}$ denote a transpose, a column vector, and a row vector, respectively.

Substituting Eq. (1.40) into Eq. (1.38), we obtain

$$F_e = \frac{1}{2} \sum_i \sum_j [\phi_i (K_{ij} - k^2 M_{ij}) \phi_j] - \sum_i \phi_i \psi_i \qquad (1.42)$$

with

$$K_{ij} = \iiint_e (\nabla N_i) \cdot (\nabla N_j) d\Omega \qquad (1.43)$$

$$M_{ij} = \iiint_e N_i N_j \, d\Omega \qquad (1.44)$$

$$\psi_i = \iint_e N_i \psi \, d\Gamma . \qquad (1.45)$$

We can express Eq. (1.42) in matrix form as

$$F_e = \frac{1}{2}\{\phi\}_e^{\mathrm{T}}[K]_e\{\phi\}_e - \frac{1}{2}k^2\{\phi\}_e^{\mathrm{T}}[M]_e\{\phi\}_e - \{\phi\}_e^{\mathrm{T}}\{\psi\}_e \qquad (1.46)$$

where the components of the $[K]_e$ and $[M]_e$ matrices are K_{ij} and M_{ij}, respectively.

Assuming that boundary conditions at the interface between the two adjacent elements are $\phi_1 = \phi_2$ and $\psi_1 = -\psi_2$, the functional for the whole region is given by

$$F = \sum_e F_e = \frac{1}{2}\{\phi\}^{\mathrm{T}}[A]\{\phi\} - \{\phi\}^{\mathrm{T}}\{\psi\} \qquad (1.47)$$

with

$$[A] = [K] - k^2[M] \qquad (1.48)$$

where the components of the $\{\phi\}$ vector are the values of ϕ at all nodes, and the global matrices $[K]$ and $[M]$ come from adding the element matrices, $[K]_e$ and $[M]_e$, respectively. The components of the $\{\psi\}$ vector corresponding to nodes on the boundaries Γ_n and Γ_f are $\hat{\psi}$ and unknown, respectively, and the other components become zero.

The first variation δF is given by

$$\delta F = \frac{1}{2}\delta\{\phi\}^{\mathrm{T}}[A]\{\phi\} + \frac{1}{2}\{\phi\}^{\mathrm{T}}[A]\delta\{\phi\} - \delta\{\phi\}^{\mathrm{T}}\{\psi\} \qquad (1.49)$$

where $\delta\{\phi\}$ is a small admissible variation of $\{\phi\}$.

As the matrix $[A]$ is symmetric, we have

$$\delta\{\phi\}^{\mathrm{T}}[A]\{\phi\} = \{\phi\}^{\mathrm{T}}[A]\delta\{\phi\}. \qquad (1.50)$$

Substituting Eq. (1.50) into Eq. (1.49), δF becomes

$$\delta F = \delta\{\phi\}^{\mathrm{T}}[A]\{\phi\} - \delta\{\phi\}^{\mathrm{T}}\{\psi\}. \qquad (1.51)$$

Using the variational principle, Eq. (1.37), the discretized algebraic equation is derived as

$$[A]\{\phi\} = \{\psi\}. \qquad (1.52)$$

For free vibration problems without the excitation term $\{\psi\}$, we obtain

$$[K]\{\phi\} - k^2[M]\{\phi\} = \{0\} \qquad (1.53)$$

where $\{0\}$ is a null vector. The so-called eigenvalue problem, Eq. (1.53), can be solved by using computer programs for generalized eigenvalue problems, and thus we obtain eigenvalues for the system under consideration, namely, eigen frequencies.

In this section Eq. (1.52) was derived by applying a variational principle to the functional, Eq. (1.36). It should be noted that the Galerkin method will yield the identical equation to Eq. (1.52) derived from a variational principle. The variational expression, Eq. (1.36), is not suitable for the study of dissipative systems, and the use of the Galerkin procedure is recommended.

1.4 Elements

Various elements are available in the FEM. However, in this section, we briefly summarize only the elements that are widely used in optical waveguide problems.

To evaluate element matrices we note that two transformations are necessary. In the first place as the shape function N_i is defined in terms of local coordinates it is necessary to devise some means of expressing the global derivatives of the type occurring in Eq. (1.43) in terms of local derivatives. In the second place the element of volume over which the integration has to be carried out needs to be expressed in terms of the local coordinates with an appropriate change of limits of integration. This section also gives the coordinate transformation relation between the local and global coordinates.

1.4.1 Coordinate transformation

Consider the set of local coordinates ξ, η, ζ and a corresponding set of global coordinates x, y, z as follows:

$$x = x(\xi, \eta, \zeta) \tag{1.54a}$$
$$y = y(\xi, \eta, \zeta) \tag{1.54b}$$
$$z = z(\xi, \eta, \zeta) . \tag{1.54c}$$

By the usual rules of partial differentiation we can write the transformation relation for differentiation as

$$\begin{bmatrix} \partial/\partial\xi \\ \partial/\partial\eta \\ \partial/\partial\zeta \end{bmatrix} = [J] \begin{bmatrix} \partial/\partial x \\ \partial/\partial y \\ \partial/\partial z \end{bmatrix} \tag{1.55}$$

where the matrix $[J]$ is known as the Jacobian matrix and is given by

$$[J] = \begin{bmatrix} \partial x/\partial\xi & \partial y/\partial\xi & \partial z/\partial\xi \\ \partial x/\partial\eta & \partial y/\partial\eta & \partial z/\partial\eta \\ \partial x/\partial\zeta & \partial y/\partial\zeta & \partial z/\partial\zeta \end{bmatrix} . \tag{1.56}$$

To find now the global derivatives we invert $[J]$ and write

$$\begin{bmatrix} \partial/\partial x \\ \partial/\partial y \\ \partial/\partial z \end{bmatrix} = [J]^{-1} \begin{bmatrix} \partial/\partial \xi \\ \partial/\partial \eta \\ \partial/\partial \zeta \end{bmatrix} \tag{1.57}$$

where the matrix $[J]^{-1}$ is the inverse of the Jacobian matrix. The transformation relation for integration becomes

$$\iiint f(x, y, z)\, dx\, dy\, dz = \iiint f(\xi, \eta, \zeta) |J(\xi, \eta, \zeta)|\, d\xi\, d\eta\, d\zeta \tag{1.58}$$

where $|J|$ is the determination of the Jacobian matrix and is known as the Jacobian.

1.4.2 Line elements

In the one-dimensional problem, line elements as shown in Fig. 1.2 are used and line coordinates L_1, L_2 are introduced. The relation equation between the line coordinates and Cartesian coordinates is given by

$$\begin{bmatrix} 1 \\ x \end{bmatrix} = \begin{bmatrix} 1 & 1 \\ x_1 & x_2 \end{bmatrix} \begin{bmatrix} L_1 \\ L_2 \end{bmatrix} \tag{1.59}$$

or

$$\begin{bmatrix} L_1 \\ L_2 \end{bmatrix} = \frac{1}{l_e} \begin{bmatrix} x_2 & -1 \\ -x_1 & 1 \end{bmatrix} \begin{bmatrix} 1 \\ x \end{bmatrix} \tag{1.60}$$

where x_k is the Cartesian coordinate of the edge k $(k = 1, 2)$ of the line and the length of the element, l_e, is given by

$$l_e = x_2 - x_1 \,. \tag{1.61}$$

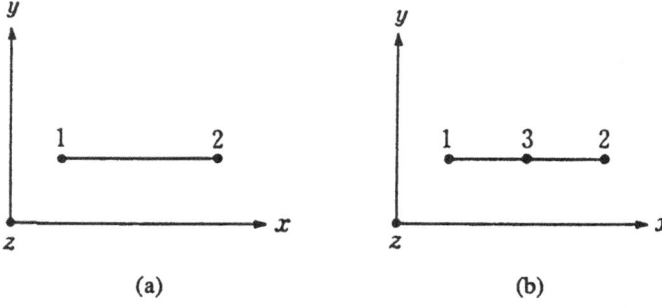

Figure 1.2: Line elements (one dimensional). (a) Linear element. (b) Quadratic element.

Defining the local coordinate ξ as

$$
\begin{aligned}
L_1 &= \xi & \text{(1.62a)} \\
L_2 &= 1 - L_1 = 1 - \xi & \text{(1.62b)}
\end{aligned}
$$

the transformation relation for differentiation is given by

$$
\frac{d}{d\xi} = -l_e \frac{d}{dx} \tag{1.63}
$$

or

$$
\frac{d}{dx} = -\frac{1}{l_e} \frac{d}{d\xi} . \tag{1.64}
$$

The transformation relation for integration is given by

$$
\int_e f(x)\, dx = l_e \int_0^1 f(\xi)\, d\xi . \tag{1.65}
$$

From Eqs. (1.62) to (1.65) we can write differentiation and integratiɔn formulas as

$$
\begin{aligned}
\frac{df}{dx} &= -\frac{1}{l_e} \left(\frac{\partial f}{\partial L_1} \frac{\partial L_1}{\partial \xi} + \frac{\partial f}{\partial L_2} \frac{\partial L_2}{\partial \xi} \right) \\
&= \frac{1}{l_e} \left(-\frac{\partial f}{\partial L_1} + \frac{\partial f}{\partial L_2} \right) \tag{1.66}
\end{aligned}
$$

$$
\begin{aligned}
\int_e L_1^k L_2^l\, dx &= l_e \int_0^1 \xi^k (1 - \xi)^l\, d\xi \\
&= l_e \frac{k!\, l!}{(k + l + 1)!} . \tag{1.67}
\end{aligned}
$$

The shape function vector and its derivative are summarized in Table 1.1.

From the above relations the integrals necessary to construct element matrices are calculated as follows:

(1) Linear elements:

$$
\int_e \{N\}\{N\}^{\mathrm{T}}\, dx = \frac{l_e}{6} \begin{bmatrix} 2 & 1 \\ 1 & 2 \end{bmatrix} \tag{1.68}
$$

$$
\int_e \{N_x\}\{N_x\}^{\mathrm{T}}\, dx = \frac{1}{l_e} \begin{bmatrix} 1 & -1 \\ -1 & 1 \end{bmatrix} \tag{1.69}
$$

$$
\int_e \{N_x\}\{N\}^{\mathrm{T}}\, dx = \frac{1}{2} \begin{bmatrix} -1 & -1 \\ 1 & 1 \end{bmatrix} . \tag{1.70}
$$

Table 1.1: Shape function for line elements and its derivative.

Element	Node number	(L_1, L_2)	$\{N\}$	$\{N_x\} = \dfrac{d\{N\}}{dx}$
Linear	1 2	$(1,0)$ $(0,1)$	$\begin{bmatrix} L_1 \\ L_2 \end{bmatrix}$	$\dfrac{1}{l_e}\begin{bmatrix} -1 \\ 1 \end{bmatrix}$
Quadratic	1 2 3	$(1,0)$ $(0,1)$ $(\frac{1}{2},\frac{1}{2})$	$\begin{bmatrix} L_1(2L_1-1) \\ L_2(2L_2-1) \\ 4L_1L_2 \end{bmatrix}$	$\dfrac{1}{l_e}\begin{bmatrix} 1-4L_1 \\ 4L_2-1 \\ 4(L_1-L_2) \end{bmatrix}$

(2) Quadratic elements:

$$\int_e \{N\}\{N\}^T \, dx = \frac{l_e}{30}\begin{bmatrix} 4 & -1 & 2 \\ -1 & 4 & 2 \\ 2 & 2 & 16 \end{bmatrix} \tag{1.71}$$

$$\int_e \{N_x\}\{N_x\}^T \, dx = \frac{1}{3l_e}\begin{bmatrix} 7 & 1 & -8 \\ 1 & 7 & -8 \\ -8 & -8 & 16 \end{bmatrix} \tag{1.72}$$

$$\int_e \{N_x\}\{N\}^T \, dx = \frac{1}{6}\begin{bmatrix} -3 & 1 & -4 \\ -1 & 3 & 4 \\ 4 & -4 & 0 \end{bmatrix}. \tag{1.73}$$

Note that the matrix $\{N\}\{N_x\}^T$ is the transpose of the matrix $\{N_x\}\{N\}^T$.

1.4.3   Triangular elements

In the two-dimensional problem triangular elements as shown in Fig. 1.3 are used and area coordinates L_1, L_2, L_3 are introduced. The relation equation between the area coordinates and Cartesian coordinates is given by

$$\begin{bmatrix} 1 \\ x \\ y \end{bmatrix} = \begin{bmatrix} 1 & 1 & 1 \\ x_1 & x_2 & x_3 \\ y_1 & y_2 & y_3 \end{bmatrix}\begin{bmatrix} L_1 \\ L_2 \\ L_3 \end{bmatrix} \tag{1.74}$$

or

$$\begin{bmatrix} L_1 \\ L_2 \\ L_3 \end{bmatrix} = \begin{bmatrix} 1 & 1 & 1 \\ x_1 & x_2 & x_3 \\ y_1 & y_2 & y_3 \end{bmatrix}^{-1}\begin{bmatrix} 1 \\ x \\ y \end{bmatrix} = \frac{1}{2A_e}\begin{bmatrix} a_1 & b_1 & c_1 \\ a_2 & b_2 & c_2 \\ a_3 & b_3 & c_3 \end{bmatrix}\begin{bmatrix} 1 \\ x \\ y \end{bmatrix} \tag{1.75}$$

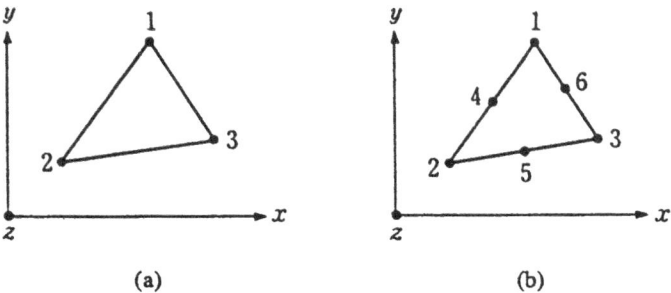

(a)   (b)

Figure 1.3: Triangular elements (two dimensional). (a) Linear element. (b) Quadratic element.

where x_k, y_k are the Cartesian coordinates of the vertex k $(k = 1, 2, 3)$ of the triangle, and the area of the element, A_e, and the coefficients a_k, b_k, c_k are given by

$$2A_e = \begin{vmatrix} 1 & 1 & 1 \\ x_1 & x_2 & x_3 \\ y_1 & y_2 & y_3 \end{vmatrix} \qquad (1.76)$$

$$a_k = x_l y_m - x_m y_l \qquad (1.77)$$

$$b_k = y_l - y_m \qquad (1.78)$$

$$c_k = x_m - x_l . \qquad (1.79)$$

Here the subscripts k, l, m always progress modulo 3, i.e., cyclically around the three vertices of the triangle.

Defining the local coordinates ξ, η as

$$L_1 = \xi \qquad (1.80a)$$

$$L_2 = \eta \qquad (1.80b)$$

$$L_3 = 1 - L_1 - L_2 = 1 - \xi - \eta \qquad (1.80c)$$

the transformation relation for differentiation is given by

$$\begin{bmatrix} \partial/\partial\xi \\ \partial/\partial\eta \end{bmatrix} = [J] \begin{bmatrix} \partial/\partial x \\ \partial/\partial y \end{bmatrix} \qquad (1.81)$$

with

$$[J] = \begin{bmatrix} x_1 - x_3 & y_1 - y_3 \\ x_2 - x_3 & y_2 - y_3 \end{bmatrix} \qquad (1.82)$$

or

$$\begin{bmatrix} \partial/\partial x \\ \partial/\partial y \end{bmatrix} = [J]^{-1} \begin{bmatrix} \partial/\partial\xi \\ \partial/\partial\eta \end{bmatrix} \qquad (1.83)$$

with

$$[J]^{-1} = \frac{1}{2A_e} \begin{bmatrix} b_1 & b_2 \\ c_1 & c_2 \end{bmatrix} . \tag{1.84}$$

The transformation relation for integration is given by

$$\iint_e f(x,y)\, dx\, dy = 2A_e \int_0^1 \int_0^{1-\xi} f(\xi,\eta)\, d\xi\, d\eta . \tag{1.85}$$

From Eqs. (1.80) to (1.85) we can write differentiation and integration formulas as

$$\frac{\partial f}{\partial x} = \frac{1}{2A_e}\left(b_1\frac{\partial f}{\partial L_1} + b_2\frac{\partial f}{\partial L_2} + b_3\frac{\partial f}{\partial L_3}\right) \tag{1.86a}$$

$$\frac{\partial f}{\partial y} = \frac{1}{2A_e}\left(c_1\frac{\partial f}{\partial L_1} + c_2\frac{\partial f}{\partial L_2} + c_3\frac{\partial f}{\partial L_3}\right) \tag{1.86b}$$

$$\iint_e L_1^k L_2^l L_3^m\, dx\, dy$$

$$= 2A_e \int_0^1 \xi^k\left[\int_0^{1-\xi} \eta^l(1-\xi-\eta)^m\, d\eta\right] d\xi$$

$$= 2A_e \frac{k!\, l!\, m!}{(k+l+m+2)!} . \tag{1.87}$$

The shape function vector and its derivative are summarized in Table 1.2.

From the above relations the integrals necessary to construct element matrices are calculated as follows:

(1) Linear elements:

$$\iint_e \{N\}\{N\}^{\mathrm{T}}\, dx\, dy = \frac{A_e}{12}\begin{bmatrix} 2 & 1 & 1 \\ 1 & 2 & 1 \\ 1 & 1 & 2 \end{bmatrix} \tag{1.88}$$

$$\left[\iint_e \{N_\mu\}\{N_\nu\}^{\mathrm{T}}\, dx\, dy\right]_{ij} = \iint_e N_{\mu i} N_{\nu j}\, dx\, dy$$

$$= A_e C_{\mu i} C_{\nu j} \quad \mu,\nu = x,y \tag{1.89}$$

$$\left[\iint_e \{N_\mu\}\{N\}^{\mathrm{T}}\, dx\, dy\right]_{ij} = \iint_e N_{\mu i} N_j\, dx\, dy$$

$$= \frac{A_e}{3} C_{\mu i} \quad \mu = x,y \tag{1.90}$$

where $[\,\cdot\,]_{ij}$ indicates the (i,j) component of the matrix $[\,\cdot\,]$ and the values of C_{xi} and C_{yi} are listed in Table 1.3.

Table 1.2: Shape function for triangular elements and its derivative.

Element	Node number	(L_1, L_2, L_3)	$\{N\}$	$\{N_x\} = \dfrac{\partial\{N\}}{\partial x}$	$\{N_y\} = \dfrac{\partial\{N\}}{\partial y}$
Linear	1 2 3	$(1,0,0)$ $(0,1,0)$ $(0,0,1)$	$\begin{bmatrix} L_1 \\ L_2 \\ L_3 \end{bmatrix}$	$\dfrac{1}{2A_e}\begin{bmatrix} b_1 \\ b_2 \\ b_3 \end{bmatrix}$	$\dfrac{1}{2A_e}\begin{bmatrix} c_1 \\ c_2 \\ c_3 \end{bmatrix}$
Quadratic	1 2 3 4 5 6	$(1,0,0)$ $(0,1,0)$ $(0,0,1)$ $(\frac{1}{2},\frac{1}{2},0)$ $(0,\frac{1}{2},\frac{1}{2})$ $(\frac{1}{2},0,\frac{1}{2})$	$\begin{bmatrix} L_1(2L_1-1) \\ L_2(2L_2-1) \\ L_3(2L_3-1) \\ 4L_1L_2 \\ 4L_2L_3 \\ 4L_3L_1 \end{bmatrix}$	$\dfrac{1}{2A_e}\begin{bmatrix} b_1(4L_1-1) \\ b_2(4L_2-1) \\ b_3(4L_3-1) \\ 4(b_1L_2+b_2L_1) \\ 4(b_2L_3+b_3L_2) \\ 4(b_3L_1+b_1L_3) \end{bmatrix}$	$\dfrac{1}{2A_e}\begin{bmatrix} c_1(4L_1-1) \\ c_2(4L_2-1) \\ c_3(4L_3-1) \\ 4(c_1L_2+c_2L_1) \\ 4(c_2L_3+c_3L_2) \\ 4(c_3L_1+c_1L_3) \end{bmatrix}$

Table 1.3: Values of C_{xi} and C_{yi}.

i	C_{xi}	C_{yi}
1	b_1	c_1
2	b_2	c_2
3	b_3	c_3

Common denominator: $1/2A_e$

(2) Quadratic elements:

$$\iint_e \{N\}\{N\}^T \, dx \, dy = \frac{A_e}{180} \begin{bmatrix} 6 & -1 & -1 & 0 & -4 & 0 \\ -1 & 6 & -1 & 0 & 0 & -4 \\ -1 & -1 & 6 & -4 & 0 & 0 \\ 0 & 0 & -4 & 32 & 16 & 16 \\ -4 & 0 & 0 & 16 & 32 & 16 \\ 0 & -4 & 0 & 16 & 16 & 32 \end{bmatrix}$$

$$(1.91)$$

$$\left[\iint_e \{N_\mu\}\{N_\nu\}^T \, dx \, dy \right]_{ij}$$

$$= \iint_e N_{\mu i} N_{\nu j} \, dx \, dy$$

$$= \frac{A_e}{6} (C_{\mu i}^{(1)} C_{\nu j}^{(1)} + C_{\mu i}^{(2)} C_{\nu j}^{(2)} + C_{\mu i}^{(3)} C_{\nu j}^{(3)})$$

$$+ \frac{A_e}{12} (C_{\mu i}^{(1)} C_{\nu j}^{(2)} + C_{\mu i}^{(1)} C_{\nu j}^{(3)} + C_{\mu i}^{(2)} C_{\nu j}^{(1)}$$

$$+ C_{\mu i}^{(2)} C_{\nu j}^{(3)} + C_{\mu i}^{(3)} C_{\nu j}^{(1)} + C_{\mu i}^{(3)} C_{\nu j}^{(2)})$$

$$+ \frac{A_e}{3} (C_{\mu i}^{(1)} C_{\nu j}^{(4)} + C_{\mu i}^{(2)} C_{\nu j}^{(4)} + C_{\mu i}^{(3)} C_{\nu j}^{(4)}$$

$$+ C_{\mu i}^{(4)} C_{\nu j}^{(1)} + C_{\mu i}^{(4)} C_{\nu j}^{(2)} + C_{\mu i}^{(4)} C_{\nu j}^{(3)})$$

$$+ A_e C_{\mu i}^{(4)} C_{\nu j}^{(4)} \qquad \mu, \nu = x, y \qquad (1.92)$$

$$\left[\iint_e \{N_\mu\}\{N\}^T \, dx \, dy \right]_{i1}$$

$$= \iint_e N_{\mu i} N_1 \, dx \, dy$$

$$= \frac{A_e}{30} C_{\mu i}^{(1)} - \frac{A_e}{60} (C_{\mu i}^{(2)} + C_{\mu i}^{(3)}) \qquad \mu = x, y \tag{1.93}$$

$$\left[\iint_e \{N_\mu\}\{N\}^T \, dx \, dy \right]_{i2}$$

$$= \iint_e N_{\mu i} N_2 \, dx \, dy$$

$$= \frac{A_e}{30} C_{\mu i}^{(2)} - \frac{A_e}{60} (C_{\mu i}^{(1)} + C_{\mu i}^{(3)}) \qquad \mu = x, y \tag{1.94}$$

$$\left[\iint_e \{N_\mu\}\{N\}^T \, dx \, dy \right]_{i3}$$

$$= \iint_e N_{\mu i} N_3 \, dx \, dy$$

$$= \frac{A_e}{30} C_{\mu i}^{(3)} - \frac{A_e}{60} (C_{\mu i}^{(1)} + C_{\mu i}^{(2)}) \qquad \mu = x, y \tag{1.95}$$

$$\left[\iint_e \{N_\mu\}\{N\}^T \, dx \, dy \right]_{i4}$$

$$= \iint_e N_{\mu i} N_4 \, dx \, dy$$

$$= \frac{A_e}{15} (2C_{\mu i}^{(1)} + 2C_{\mu i}^{(2)} + C_{\mu i}^{(3)} + 5C_{\mu i}^{(4)}) \qquad \mu = x, y \tag{1.96}$$

$$\left[\iint_e \{N_\mu\}\{N\}^T \, dx \, dy \right]_{i5}$$

$$= \iint_e N_{\mu i} N_5 \, dx \, dy$$

$$= \frac{A_e}{15} (C_{\mu i}^{(1)} + 2C_{\mu i}^{(2)} + 2C_{\mu i}^{(3)} + 5C_{\mu i}^{(4)}) \qquad \mu = x, y \tag{1.97}$$

$$\left[\iint_e \{N_\mu\}\{N\}^T \, dx \, dy \right]_{i6}$$

$$= \iint_e N_{\mu i} N_6 \, dx \, dy$$

$$= \frac{A_e}{15} (2C_{\mu i}^{(1)} + C_{\mu i}^{(2)} + 2C_{\mu i}^{(3)} + 5C_{\mu i}^{(4)}) \qquad \mu = x, y \tag{1.98}$$

where the values of $C_{xi}^{(1)}$ to $C_{yi}^{(4)}$ are listed in Table 1.4.

1.4.4 Tetrahedral elements

In the three-dimensional problem tetrahedral elements as shown in Fig. 1.4 are used and volume coordinates L_1, L_2, L_3, L_4 are introduced.

Table 1.4: Values of $C_{xi}^{(1)}$ to $C_{yi}^{(4)}$.

i	$C_{xi}^{(1)}$	$C_{xi}^{(2)}$	$C_{xi}^{(3)}$	$C_{xi}^{(4)}$	$C_{yi}^{(1)}$	$C_{yi}^{(2)}$	$C_{yi}^{(3)}$	$C_{yi}^{(4)}$
1	$4b_1$			$-b_1$	$4c_1$			$-c_1$
2		$4b_2$		$-b_2$		$4c_2$		$-c_2$
3			$4b_3$	$-b_3$			$4c_3$	$-c_3$
4	$4b_2$	$4b_1$			$4c_2$	$4c_1$		
5		$4b_3$	$4b_2$			$4c_3$	$4c_2$	
6	$4b_3$		$4b_1$		$4c_3$		$4c_1$	

Common denominator: $1/2A_e$

The relation equation between the volume coordinates and Cartesian coordinates is given by

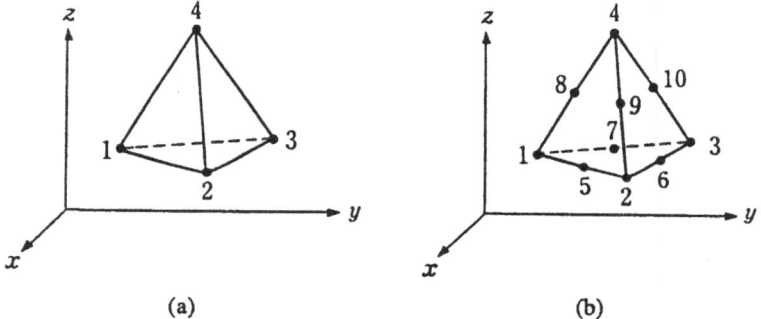

Figure 1.4: Tetrahedral elements (three dimensional). (a) Linear element. (b) Quadratic element.

$$\begin{bmatrix} 1 \\ x \\ y \\ z \end{bmatrix} = \begin{bmatrix} 1 & 1 & 1 & 1 \\ x_1 & x_2 & x_3 & x_4 \\ y_1 & y_2 & y_3 & y_4 \\ z_1 & z_2 & z_3 & z_4 \end{bmatrix} \begin{bmatrix} L_1 \\ L_2 \\ L_3 \\ L_4 \end{bmatrix} \tag{1.99}$$

or

$$\begin{bmatrix} L_1 \\ L_2 \\ L_3 \\ L_4 \end{bmatrix} = \begin{bmatrix} 1 & 1 & 1 & 1 \\ x_1 & x_2 & x_3 & x_4 \\ y_1 & y_2 & y_3 & y_4 \\ z_1 & z_2 & z_3 & z_4 \end{bmatrix}^{-1} \begin{bmatrix} 1 \\ x \\ y \\ z \end{bmatrix} = \frac{1}{6V_e} \begin{bmatrix} a_1 & b_1 & c_1 & d_1 \\ a_2 & b_2 & c_2 & d_2 \\ a_3 & b_3 & c_3 & d_3 \\ a_4 & b_4 & c_4 & d_4 \end{bmatrix} \begin{bmatrix} 1 \\ x \\ y \\ z \end{bmatrix} \tag{1.100}$$

where x_k, y_k, z_k are the Cartesian coordinates of the vertex k ($k = 1, 2, 3, 4$) of the tetrahedron, and the volume of the element, V_e, and the coefficients a_k, b_k, c_k, d_k are given by

$$6V_e = \begin{vmatrix} 1 & 1 & 1 & 1 \\ x_1 & x_2 & x_3 & x_4 \\ y_1 & y_2 & y_3 & y_4 \\ z_1 & z_2 & z_3 & z_4 \end{vmatrix} \tag{1.101}$$

$$\begin{aligned} a_k &= \alpha_k[x_l(y_m z_n - y_n z_m) + x_m(y_n z_l - y_l z_n) \\ &\quad + x_n(y_l z_m - y_m z_l)] \end{aligned} \tag{1.102}$$

$$b_k = \alpha_k[y_l(z_n - z_m) + y_m(z_l - z_n) + y_n(z_m - z_l)] \tag{1.103}$$

$$c_k = \alpha_k[z_l(x_n - x_m) + z_m(x_l - x_n) + z_n(x_m - x_l)] \tag{1.104}$$

$$d_k = \alpha_k[x_l(y_n - y_m) + x_m(y_l - y_n) + x_n(y_m - y_l)] \tag{1.105}$$

with

$$\alpha_k = \begin{cases} 1 & \text{for } k=1, 3 \\ -1 & \text{for } k=2, 4. \end{cases} \tag{1.106}$$

Here the subscripts k, l, m, n always progress modulo 4, i.e., cyclically around the four vertices of the tetrahedron.

Defining the local coordinates ξ, η, ζ as

$$L_1 = \xi \tag{1.107a}$$
$$L_2 = \eta \tag{1.107b}$$
$$L_3 = \zeta \tag{1.107c}$$
$$L_4 = 1 - L_1 - L_2 - L_3 = 1 - \xi - \eta - \zeta \tag{1.107d}$$

the transformation relation for differentiation is given by

$$\begin{bmatrix} \partial/\partial\xi \\ \partial/\partial\eta \\ \partial/\partial\zeta \end{bmatrix} = [J] \begin{bmatrix} \partial/\partial x \\ \partial/\partial y \\ \partial/\partial z \end{bmatrix} \tag{1.108}$$

with

$$[J] = \begin{bmatrix} x_1 - x_4 & y_1 - y_4 & z_1 - z_4 \\ x_2 - x_4 & y_2 - y_4 & z_2 - z_4 \\ x_3 - x_4 & y_3 - y_4 & z_3 - z_4 \end{bmatrix} \tag{1.109}$$

or

$$\begin{bmatrix} \partial/\partial x \\ \partial/\partial y \\ \partial/\partial z \end{bmatrix} = [J]^{-1} \begin{bmatrix} \partial/\partial\xi \\ \partial/\partial\eta \\ \partial/\partial\zeta \end{bmatrix} \tag{1.110}$$

with

$$[J]^{-1} = \frac{1}{6V_e} \begin{bmatrix} b_1 & b_2 & b_3 \\ c_1 & c_2 & c_3 \\ d_1 & d_2 & d_3 \end{bmatrix}. \tag{1.111}$$

The transformation relation for integration is given by

$$\iiint_e f(x,y,z)\, dx\, dy\, dz = 6V_e \int_0^1 \int_0^{1-\xi} \int_0^{1-\xi-\eta} f(\xi,\eta,\zeta)\, d\xi\, d\eta\, d\zeta. \tag{1.112}$$

From Eqs. (1.107) to (1.112) we can write differentiation and integration formulas as

$$\frac{\partial f}{\partial x} = \frac{1}{6V_e}\left(b_1\frac{\partial f}{\partial L_1} + b_2\frac{\partial f}{\partial L_2} + b_3\frac{\partial f}{\partial L_3} + b_4\frac{\partial f}{\partial L_4}\right) \tag{1.113a}$$

$$\frac{\partial f}{\partial y} = \frac{1}{6V_e}\left(c_1\frac{\partial f}{\partial L_1} + c_2\frac{\partial f}{\partial L_2} + c_3\frac{\partial f}{\partial L_3} + c_4\frac{\partial f}{\partial L_4}\right) \tag{1.113b}$$

$$\frac{\partial f}{\partial z} = \frac{1}{6V_e}\left(d_1\frac{\partial f}{\partial L_1} + d_2\frac{\partial f}{\partial L_2} + d_3\frac{\partial f}{\partial L_3} + d_4\frac{\partial f}{\partial L_4}\right) \tag{1.113c}$$

$$\iiint_e L_1^k L_2^l L_3^m L_4^n\, dx\, dy\, dz$$

$$= 6V_e \int_0^1 \xi^k \left\{\int_0^{1-\xi} \eta^l \left[\int_0^{1-\xi-\eta} \zeta^m (1-\xi-\eta-\zeta)^n\, d\zeta\right] d\eta\right\} d\xi$$

$$= 6V_e \frac{k!\, l!\, m!\, n!}{(k+l+m+n+3)!}. \tag{1.114}$$

The shape function vector and its derivative are summarized in Table 1.5.

From the above relations the integrals necessary to construct element matrices are calculated as follows:

(1) Linear elements:

$$\iiint_e \{N\}\{N\}^{\mathrm{T}}\, dx\, dy\, dz = \frac{V_e}{20} \begin{bmatrix} 2 & 1 & 1 & 1 \\ 1 & 2 & 1 & 1 \\ 1 & 1 & 2 & 1 \\ 1 & 1 & 1 & 2 \end{bmatrix} \tag{1.115}$$

$$\left[\iiint_e \{N_\mu\}\{N_\nu\}^{\mathrm{T}}\, dx\, dy\, dz\right]_{ij}$$

$$= \iiint_e N_{\mu i} N_{\nu j}\, dx\, dy\, dz$$

$$= V_e C_{\mu i} C_{\nu j} \qquad \mu,\nu = x,y,z \tag{1.116}$$

Table 1.5: Shape function for tetrahedral elements and its derivative.

Element	Node number	(L_1, L_2, L_3, L_4)	$\{N\}$	$\{N_x\} = \frac{\partial\{N\}}{\partial x}$	$\{N_y\} = \frac{\partial\{N\}}{\partial y}$	$\{N_z\} = \frac{\partial\{N\}}{\partial z}$
Linear	1	$(1,0,0,0)$	L_1	$\frac{1}{6V_e}\,b_1$	$\frac{1}{6V_e}\,c_1$	$\frac{1}{6V_e}\,d_1$
	2	$(0,1,0,0)$	L_2	b_2	c_2	d_2
	3	$(0,0,1,0)$	L_3	b_3	c_3	d_3
	4	$(0,0,0,1)$	L_4	b_4	c_4	d_4
Quadratic	1	$(1,0,0,0)$	$L_1(2L_1-1)$	$\frac{1}{6V_e}\,b_1(4L_1-1)$	$\frac{1}{6V_e}\,c_1(4L_1-1)$	$\frac{1}{6V_e}\,d_1(4L_1-1)$
	2	$(0,1,0,0)$	$L_2(2L_2-1)$	$b_2(4L_2-1)$	$c_2(4L_2-1)$	$d_2(4L_2-1)$
	3	$(0,0,1,0)$	$L_3(2L_3-1)$	$b_3(4L_3-1)$	$c_3(4L_3-1)$	$d_3(4L_3-1)$
	4	$(0,0,0,1)$	$L_4(2L_4-1)$	$b_4(4L_4-1)$	$c_4(4L_4-1)$	$d_4(4L_4-1)$
	5	$(\frac{1}{2},\frac{1}{2},0,0)$	$4L_1L_2$	$4(b_1L_2+b_2L_1)$	$4(c_1L_2+c_2L_1)$	$4(d_1L_2+d_2L_1)$
	6	$(0,\frac{1}{2},\frac{1}{2},0)$	$4L_2L_3$	$4(b_2L_3+b_3L_2)$	$4(c_2L_3+c_3L_2)$	$4(d_2L_3+d_3L_2)$
	7	$(\frac{1}{2},0,\frac{1}{2},0)$	$4L_3L_1$	$4(b_3L_1+b_1L_3)$	$4(c_3L_1+c_1L_3)$	$4(d_3L_1+d_1L_3)$
	8	$(\frac{1}{2},0,0,\frac{1}{2})$	$4L_1L_4$	$4(b_1L_4+b_4L_1)$	$4(c_1L_4+c_4L_1)$	$4(d_1L_4+d_4L_1)$
	9	$(0,\frac{1}{2},0,\frac{1}{2})$	$4L_2L_4$	$4(b_2L_4+b_4L_2)$	$4(c_2L_4+c_4L_2)$	$4(d_2L_4+d_4L_2)$
	10	$(0,0,\frac{1}{2},\frac{1}{2})$	$4L_3L_4$	$4(b_3L_4+b_4L_3)$	$4(c_3L_4+c_4L_3)$	$4(d_3L_4+d_4L_3)$

Table 1.6: Values of C_{xi}, C_{yi}, and C_{zi}.

i	C_{xi}	C_{yi}	C_{zi}
1	b_1	c_1	d_1
2	b_2	c_2	d_2
3	b_3	c_3	d_3
4	b_4	c_4	d_4

Common denominator: $1/6V_e$

where the values of C_{xi}, C_{yi}, and C_{zi} are listed in Table 1.6.

(2) Quadratic elements:

$$\iiint_e \{N\}\{N\}^{\mathrm{T}}\, dx\, dy\, dz$$

$$= \frac{V_e}{420}
\begin{bmatrix}
6 & 1 & 1 & 1 & -4 & -6 & -4 & -4 & -6 & -6 \\
1 & 6 & 1 & 1 & -4 & -4 & -6 & -6 & -4 & -6 \\
1 & 1 & 6 & 1 & -6 & -4 & -4 & -6 & -6 & -4 \\
1 & 1 & 1 & 6 & -6 & -6 & -6 & -4 & -4 & -4 \\
-4 & -4 & -6 & -6 & 32 & 16 & 16 & 16 & 16 & 8 \\
-6 & -4 & -4 & -6 & 16 & 32 & 16 & 8 & 16 & 16 \\
-4 & -6 & -4 & -6 & 16 & 16 & 32 & 16 & 8 & 16 \\
-4 & -6 & -6 & -4 & 16 & 8 & 16 & 32 & 16 & 16 \\
-6 & -4 & -6 & -4 & 16 & 16 & 8 & 16 & 32 & 16 \\
-6 & -6 & -4 & -4 & 8 & 16 & 16 & 16 & 16 & 32
\end{bmatrix}$$

$$(1.117)$$

$$\left[\iiint_e \{N_\mu\}\{N_\nu\}^{\mathrm{T}}\, dx\, dy\, dz\right]_{ij}$$

$$= \iiint_e N_{\mu i} N_{\nu j}\, dx\, dy\, dz$$

$$= \frac{V_e}{10}(C_{\mu i}^{(1)} C_{\nu j}^{(1)} + C_{\mu i}^{(2)} C_{\nu j}^{(2)} + C_{\mu i}^{(3)} C_{\nu j}^{(3)} + C_{\mu i}^{(4)} C_{\nu j}^{(4)})$$

$$+ \frac{V_e}{20}(C_{\mu i}^{(1)} C_{\nu j}^{(2)} + C_{\mu i}^{(1)} C_{\nu j}^{(3)} + C_{\mu i}^{(1)} C_{\nu j}^{(4)} + C_{\mu i}^{(2)} C_{\nu j}^{(1)}$$

$$+ C_{\mu i}^{(2)} C_{\nu j}^{(3)} + C_{\mu i}^{(2)} C_{\nu j}^{(4)} + C_{\mu i}^{(3)} C_{\nu j}^{(1)} + C_{\mu i}^{(3)} C_{\nu j}^{(2)}$$

$$+ C_{\mu i}^{(3)} C_{\nu j}^{(4)} + C_{\mu i}^{(4)} C_{\nu j}^{(1)} + C_{\mu i}^{(4)} C_{\nu j}^{(2)} + C_{\mu i}^{(4)} C_{\nu j}^{(3)})$$

$$+ \frac{V_e}{4}(C_{\mu i}^{(1)} C_{\nu j}^{(5)} + C_{\mu i}^{(2)} C_{\nu j}^{(5)} + C_{\mu i}^{(3)} C_{\nu j}^{(5)} + C_{\mu i}^{(4)} C_{\nu j}^{(5)}$$

$$+ C_{\mu i}^{(5)} C_{\nu j}^{(1)} + C_{\mu i}^{(5)} C_{\nu j}^{(2)} + C_{\mu i}^{(5)} C_{\nu j}^{(3)} + C_{\mu i}^{(5)} C_{\nu j}^{(4)})$$

$$+ V_e C_{\mu i}^{(5)} C_{\nu j}^{(5)} \qquad \mu, \nu = x, y, z \qquad (1.118)$$

where the values of $C_{xi}^{(1)}$ to $C_{zi}^{(5)}$ are listed in Table 1.7.

1.4.5 Ring elements

In the axisymmetric two-dimensional problem, ring elements as shown in Fig. 1.5 are used. Differential and integration formulas for ring elements are obtained by replacing x by r in those for line elements.

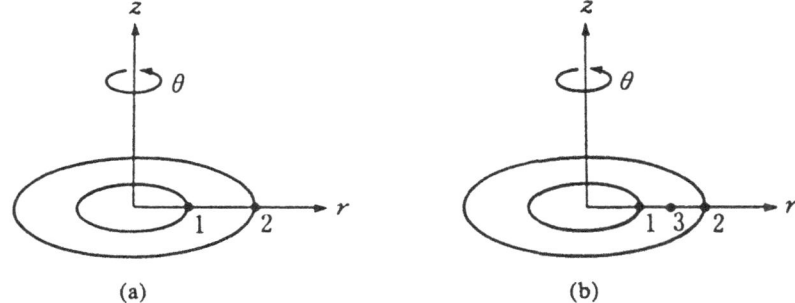

(a)   (b)

Figure 1.5: Ring elements (axisymmetric two dimensional). (a) Linear element. (b) Quadratic element.

The integrals which do not appear in the one-dimensional problem are now summarized:

(1) Linear elements:

$$\int_e r\{N\}\{N\}^{\mathrm{T}} \, dr = \frac{r_1 l_e}{12} \begin{bmatrix} 3 & 1 \\ 1 & 1 \end{bmatrix} + \frac{r_2 l_e}{12} \begin{bmatrix} 1 & 1 \\ 1 & 3 \end{bmatrix} \qquad (1.119)$$

$$\int_e r\{N_r\}\{N_r\}^{\mathrm{T}} \, dr = \frac{r_1}{2l_e} \begin{bmatrix} 1 & -1 \\ -1 & 1 \end{bmatrix} + \frac{r_2}{2l_e} \begin{bmatrix} 1 & -1 \\ -1 & 1 \end{bmatrix}$$

$$(1.120)$$

$$\int_e r\{N_r\}\{N\}^{\mathrm{T}} \, dr = \frac{r_1}{6} \begin{bmatrix} -2 & -1 \\ 2 & 1 \end{bmatrix} + \frac{r_2}{6} \begin{bmatrix} -1 & -2 \\ 1 & 2 \end{bmatrix}$$

$$(1.121)$$

$$\int_e \frac{1}{r}\{N\}\{N\}^{\mathrm{T}} \, dr = \begin{bmatrix} A_{11} & A_{12} \\ A_{12} & A_{22} \end{bmatrix} \qquad (1.122)$$

Table 1.7: Values of $C_{xi}^{(1)}$ to $C_{zi}^{(5)}$.

i	$C_{xi}^{(1)}$	$C_{xi}^{(2)}$	$C_{xi}^{(3)}$	$C_{xi}^{(4)}$	$C_{xi}^{(5)}$	$C_{yi}^{(1)}$	$C_{yi}^{(2)}$	$C_{yi}^{(3)}$	$C_{yi}^{(4)}$	$C_{yi}^{(5)}$	$C_{zi}^{(1)}$	$C_{zi}^{(2)}$	$C_{zi}^{(3)}$	$C_{zi}^{(4)}$	$C_{zi}^{(5)}$
1	$4b_1$				$-b_1$	$4c_1$				$-c_1$	$4d_1$				$-d_1$
2		$4b_2$			$-b_2$		$4c_2$			$-c_2$		$4d_2$			$-d_2$
3			$4b_3$		$-b_3$			$4c_3$		$-c_3$			$4d_3$		$-d_3$
4				$4b_4$	$-b_4$				$4c_4$	$-c_4$				$4d_4$	$-d_4$
5	$4b_2$	$4b_1$				$4c_2$	$4c_1$				$4d_2$	$4d_1$			
6		$4b_3$	$4b_2$				$4c_3$	$4c_2$				$4d_3$	$4d_2$		
7	$4b_3$		$4b_1$			$4c_3$		$4c_1$			$4d_3$		$4d_1$		
8	$4b_4$			$4b_1$		$4c_4$			$4c_1$		$4d_4$			$4d_1$	
9		$4b_4$		$4b_2$			$4c_4$		$4c_2$			$4d_4$		$4d_2$	
10			$4b_4$	$4b_3$				$4c_4$	$4c_3$				$4d_4$	$4d_3$	

Common denominator: $1/6V_e$

with

$$A_{11} = -u + 3/2 + (u+1)^2 v \qquad (1.123\text{a})$$
$$A_{12} = u + 1/2 - (u+1)uv \qquad (1.123\text{b})$$
$$A_{22} = -u + 1/2 - u^2 v \qquad (1.123\text{c})$$

where u and v are given by

$$u = r_1/l_e \qquad (1.124)$$
$$v = \ln(r_2/r_1) . \qquad (1.125)$$

When the term $1/r$ is assumed to be constant within each element, the integral, Eq. (1.122), may be calculated as

$$\int_e \frac{1}{r}\{N\}\{N\}^{\mathrm{T}}\, dr = \frac{l_e}{6r_c}\left[\begin{array}{cc} 2 & 1 \\ 1 & 2 \end{array}\right] \qquad (1.126)$$

with

$$r_c = (r_1 + r_2)/2 . \qquad (1.127)$$

(2) Quadratic elements:

$$\int_e r\{N\}\{N\}^{\mathrm{T}}\, dr$$
$$= \frac{r_1 l_e}{60}\left[\begin{array}{ccc} 7 & -1 & 4 \\ -1 & 1 & 0 \\ 4 & 0 & 16 \end{array}\right] + \frac{r_2 l_e}{60}\left[\begin{array}{ccc} 1 & -1 & 0 \\ -1 & 7 & 4 \\ 0 & 4 & 16 \end{array}\right] \qquad (1.128)$$

$$\int_e r\{N_r\}\{N_r\}^{\mathrm{T}}\, dr$$
$$= \frac{r_1}{6l_e}\left[\begin{array}{ccc} 11 & 1 & -12 \\ 1 & 3 & -4 \\ -12 & -4 & 16 \end{array}\right] + \frac{r_2}{6l_e}\left[\begin{array}{ccc} 3 & 1 & -4 \\ 1 & 11 & -12 \\ -4 & -12 & 16 \end{array}\right]$$
$$\qquad (1.129)$$

$$\int_e r\{N_r\}\{N\}^{\mathrm{T}}\, dr$$
$$= \frac{r_1}{30}\left[\begin{array}{ccc} -13 & 2 & -14 \\ -3 & 2 & 6 \\ 16 & -4 & 8 \end{array}\right] + \frac{r_2}{30}\left[\begin{array}{ccc} -2 & 3 & -6 \\ -2 & 13 & 14 \\ 4 & -16 & -8 \end{array}\right] \qquad (1.130)$$

$$\int_e \frac{1}{r}\{N\}\{N\}^{\mathrm{T}}\, dr = \left[\begin{array}{ccc} A_{11} & A_{12} & 4A_{13} \\ A_{12} & A_{22} & 4A_{23} \\ 4A_{13} & 4A_{23} & 16A_{33} \end{array}\right] \qquad (1.131)$$

with

$$A_{11} = -4u^3 - 10u^2 - (25/3)u - 5/2 + (u+1)^2(2u+1)^2 v$$
$$\tag{1.132a}$$
$$A_{12} = -4u^3 - 6u^2 - (7/3)u - 1/6 + (u+1)(2u+1)^2 uv$$
$$\tag{1.132b}$$
$$A_{13} = 2u^3 + 4u^2 + (13/6)u + 1/6 - (u+1)^2(2u+1)uv$$
$$\tag{1.132c}$$
$$A_{22} = -4u^3 - 2u^2 - (1/3)u + 1/6 + (2u+1)^2 u^2 v \qquad (1.132\mathrm{d})$$
$$A_{23} = 2u^3 + 2u^2 + (1/6)u - (u+1)(2u+1)u^2 v \qquad (1.132\mathrm{e})$$
$$A_{33} = -u^3 - (3/2)u^2 - (1/3)u + 1/12 + (u+1)^2 u^2 v \,.$$
$$\tag{1.132f}$$

When the term $1/r$ is assumed to be constant within each element, the integral, Eq. (1.131), may be calculated as

$$\int_e \frac{1}{r}\{N\}\{N\}^{\mathrm{T}}\, dr = \frac{l_e}{30 r_c} \begin{bmatrix} 4 & -1 & 2 \\ -1 & 4 & 2 \\ 2 & 2 & 16 \end{bmatrix}. \qquad (1.133)$$

Gauss-Legendre quadrature formulas (see Subsection 1.5.1) can also be utilized to calculate the integrals, Eqs. (1.122) and (1.131).

1.4.6   Triangular ring elements

In the axisymmetric three-dimensional problem, triangular ring elements as shown in Fig. 1.6 are used. Differential and integration formulas for triangular ring elements are obtained by replacing x and y by r and z in those for triangular elements, respectively.

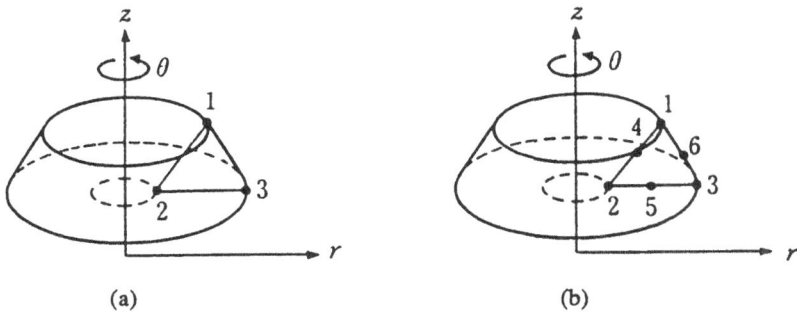

Figure 1.6: Triangular ring elements (axisymmetric three dimensional). (a) Linear element. (b) Quadratic element.

The integrals which do not appear in the two-dimensional problem are now summarized:

(1) Linear elements:

$$\iint_e r\{N\}\{N\}^{\mathrm{T}}\, dr\, dz$$

$$= \frac{r_1 A_e}{60}\begin{bmatrix} 6 & 2 & 2 \\ 2 & 2 & 1 \\ 2 & 1 & 2 \end{bmatrix} + \frac{r_2 A_e}{60}\begin{bmatrix} 2 & 2 & 1 \\ 2 & 6 & 2 \\ 1 & 2 & 2 \end{bmatrix} + \frac{r_3 A_e}{60}\begin{bmatrix} 2 & 1 & 2 \\ 1 & 2 & 2 \\ 2 & 2 & 6 \end{bmatrix}$$

$$\tag{1.134}$$

$$\left[\iint_e r\{N_\mu\}\{N_\nu\}^{\mathrm{T}}\, dr\, dz\right]_{ij}$$

$$= \iint_e r N_{\mu i} N_{\nu j}\, dr\, dz$$

$$= r_c A_e C_{\mu i} C_{\nu j} \qquad \mu,\nu = r,z \tag{1.135}$$

$$\iint_e \frac{1}{r}\{N\}\{N\}^{\mathrm{T}}\, dr\, dz = \frac{A_e}{12 r_c}\begin{bmatrix} 2 & 1 & 1 \\ 1 & 2 & 1 \\ 1 & 1 & 2 \end{bmatrix} \tag{1.136}$$

with

$$r_c = (r_1 + r_2 + r_3)/3\,. \tag{1.137}$$

(2) Quadratic elements:

$$\iint_e r\{N\}\{N\}^{\mathrm{T}}\, dr\, dz$$

$$= \frac{r_1 A_e}{1260} \begin{bmatrix} 30 & -4 & -4 & 12 & -4 & 12 \\ -4 & 6 & 1 & -8 & -4 & -12 \\ -4 & 1 & 6 & -12 & -4 & -8 \\ 12 & -8 & -12 & 96 & 32 & 48 \\ -4 & -4 & -4 & 32 & 32 & 32 \\ 12 & -12 & -8 & 48 & 32 & 96 \end{bmatrix}$$

$$+ \frac{r_2 A_e}{1260} \begin{bmatrix} 6 & -4 & 1 & -8 & -12 & -4 \\ -4 & 30 & -4 & 12 & 12 & -4 \\ 1 & -4 & 6 & -12 & -8 & -4 \\ -8 & 12 & -12 & 96 & 48 & 32 \\ -12 & 12 & -8 & 48 & 96 & 32 \\ -4 & -4 & -4 & 32 & 32 & 32 \end{bmatrix}$$

$$+ \frac{r_3 A_e}{1260} \begin{bmatrix} 6 & 1 & -4 & -4 & -12 & -8 \\ 1 & 6 & -4 & -4 & -8 & -12 \\ -4 & -4 & 30 & -4 & 12 & 12 \\ -4 & -4 & -4 & 32 & 32 & 32 \\ -12 & -8 & 12 & 32 & 96 & 48 \\ -8 & -12 & 12 & 32 & 48 & 96 \end{bmatrix} \qquad (1.138)$$

$$\left[\iint_e r \{N_\mu\}\{N_\nu\}^{\mathrm{T}} \, dr \, dz \right]_{ij}$$

$$= \iint_e r N_{\mu i} N_{\nu j} \, dr \, dz$$

$$= \frac{r_1 A_e}{30} (3C_{\mu i}^{(1)} C_{\nu j}^{(1)} + C_{\mu i}^{(2)} C_{\nu j}^{(2)} + C_{\mu i}^{(3)} C_{\nu j}^{(3)})$$

$$+ \frac{r_1 A_e}{60} (2C_{\mu i}^{(1)} C_{\nu j}^{(2)} + 2C_{\mu i}^{(1)} C_{\nu j}^{(3)} + 2C_{\mu i}^{(2)} C_{\nu j}^{(1)}$$

$$+ C_{\mu i}^{(2)} C_{\nu j}^{(3)} + 2C_{\mu i}^{(3)} C_{\nu j}^{(1)} + C_{\mu i}^{(3)} C_{\nu j}^{(2)})$$

$$+ \frac{r_1 A_e}{12} (2C_{\mu i}^{(1)} C_{\nu j}^{(4)} + C_{\mu i}^{(2)} C_{\nu j}^{(4)} + C_{\mu i}^{(3)} C_{\nu j}^{(4)}$$

$$+ 2C_{\mu i}^{(4)} C_{\nu j}^{(1)} + C_{\mu i}^{(4)} C_{\nu j}^{(2)} + C_{\mu i}^{(4)} C_{\nu j}^{(3)})$$

$$+ \frac{r_1 A_e}{3} C_{\mu i}^{(4)} C_{\nu j}^{(4)}$$

$$+ \frac{r_2 A_e}{30} (C_{\mu i}^{(1)} C_{\nu j}^{(1)} + 3C_{\mu i}^{(2)} C_{\nu j}^{(2)} + C_{\mu i}^{(3)} C_{\nu j}^{(3)})$$

$$+ \frac{r_2 A_e}{60} (2C_{\mu i}^{(1)} C_{\nu j}^{(2)} + C_{\mu i}^{(1)} C_{\nu j}^{(3)} + 2C_{\mu i}^{(2)} C_{\nu j}^{(1)}$$

$$+ 2C_{\mu i}^{(2)} C_{\nu j}^{(3)} + C_{\mu i}^{(3)} C_{\nu j}^{(1)} + 2C_{\mu i}^{(3)} C_{\nu j}^{(2)})$$

$$+ \frac{r_2 A_e}{12}(C_{\mu i}^{(1)} C_{\nu j}^{(4)} + 2C_{\mu i}^{(2)} C_{\nu j}^{(4)} + C_{\mu i}^{(3)} C_{\nu j}^{(4)}$$

$$+ C_{\mu i}^{(4)} C_{\nu j}^{(1)} + 2C_{\mu i}^{(4)} C_{\nu j}^{(2)} + C_{\mu i}^{(4)} C_{\nu j}^{(3)})$$

$$+ \frac{r_2 A_e}{3} C_{\mu i}^{(4)} C_{\nu j}^{(4)}$$

$$+ \frac{r_3 A_e}{30}(C_{\mu i}^{(1)} C_{\nu j}^{(1)} + C_{\mu i}^{(2)} C_{\nu j}^{(2)} + 3C_{\mu i}^{(3)} C_{\nu j}^{(3)})$$

$$+ \frac{r_3 A_e}{60}(C_{\mu i}^{(1)} C_{\nu j}^{(2)} + 2C_{\mu i}^{(1)} C_{\nu j}^{(3)} + C_{\mu i}^{(2)} C_{\nu j}^{(1)}$$

$$+ 2C_{\mu i}^{(2)} C_{\nu j}^{(3)} + 2C_{\mu i}^{(3)} C_{\nu j}^{(1)} + 2C_{\mu i}^{(3)} C_{\nu j}^{(2)})$$

$$+ \frac{r_3 A_e}{12}(C_{\mu i}^{(1)} C_{\nu j}^{(4)} + C_{\mu i}^{(2)} C_{\nu j}^{(4)} + 2C_{\mu i}^{(3)} C_{\nu j}^{(4)}$$

$$+ C_{\mu i}^{(4)} C_{\nu j}^{(1)} + C_{\mu i}^{(4)} C_{\nu j}^{(2)} + 2C_{\mu i}^{(4)} C_{\nu j}^{(3)})$$

$$+ \frac{r_3 A_e}{3} C_{\mu i}^{(4)} C_{\nu j}^{(4)} \qquad \mu, \nu = r, z \qquad (1.139)$$

$$\iint_e \frac{1}{r} \{N\}\{N\}^{\mathrm{T}} \, dr \, dz = \frac{A_e}{180 r_c} \begin{bmatrix} 6 & -1 & -1 & 0 & -4 & 0 \\ -1 & 6 & -1 & 0 & 0 & -4 \\ -1 & -1 & 6 & -4 & 0 & 0 \\ 0 & 0 & -4 & 32 & 16 & 16 \\ -4 & 0 & 0 & 16 & 32 & 16 \\ 0 & -4 & 0 & 16 & 16 & 32 \end{bmatrix} .$$

$$(1.140)$$

Numerical integration formulas derived by Hammer *et al.* (see Subsection 1.5.2) can also be utilized to calculate the integrals, Eqs. (1.136) and (1.140).

1.5 Numerical Integration

The FEM are always based on an integral formulation. Therefore the accuracy of the integration is quite important. In optical waveguides with graded-index profiles it is impossible or impractical to integrate the expression in closed form and numerical integration must be utilized. If one is using elements with curved or distorted sides, such as isoparametric elements (see Section 1.6), it is almost always necessary to use numerical integration. The use of numerical integration avoids lengthy algebraic expressions and simplifies the programming of the element matrices.

There are many numerical integration methods available. Only those methods commonly used in finite element applications will be summarized here together with tables of convenient numerical coefficients.

1.5.1   One-dimensional integration

Using the local coordinate ξ defined by

$$x = \frac{1}{2}(x_2 - x_1)\xi + \frac{1}{2}(x_2 + x_1) \tag{1.141}$$

one-dimensional integration may be expressed as

$$\int_{x_1}^{x_2} f(x)\,dx = \frac{1}{2}(x_2 - x_1) \int_{-1}^{1} f(\xi)\,d\xi. \tag{1.142}$$

Gauss-Legendre quadrature formula is of the form

$$\int_{-1}^{1} f(\xi)\,d\xi = \sum_{i=1}^{n} W_i f(\xi_i) \tag{1.143}$$

where W_i and ξ_i are the weighting coefficients and local coordinates associated with n sampling points, respectively. These data are presented in Table 1.8.

Table 1.8: Gauss-Legendre quadrature formulas.

n	i	W_i	ξ_i
1	1	2.0	0.0
2	1	1.0	−0.57735027
	2	1.0	0.57735027
3	1	0.55555556	−0.77459667
	2	0.88888889	0.0
	3	0.55555556	0.77459667
4	1	0.34785485	−0.86113631
	2	0.65214515	−0.33998104
	3	0.65214515	0.33998104
	4	0.34785485	0.86113631
5	1	0.23692689	−0.90617985
	2	0.47862867	−0.53846931
	3	0.56888889	0.0
	4	0.47862867	0.53846931
	5	0.23692689	0.90617985

If the line coordinates L_1, L_2 are used, define a change of variable

$$L_1 = (1 - \xi)/2 \tag{1.144a}$$
$$L_2 = (1 + \xi)/2 \tag{1.144b}$$

so that the formula, Eq. (1.143), will be utilized.

1.5.2 Two-dimensional integration

Using the local coordinates ξ, η, two-dimensional integration may be expressed as

$$\iint f(x,y)\,dx\,dy = \iint f(\xi,\eta)|J(\xi,\eta)|d\xi\,d\eta\ . \qquad (1.145)$$

For a triangle, in terms of the area coordinates (see Eq. (1.80)) the numerical integration formula derived by Hammer *et al.* is of the form

$$\iint_e f(L_1,L_2,L_3)|J(L_1,L_2,L_3)|d\xi\,d\eta$$

$$= \sum_{i=1}^{n}\frac{1}{2}W_i f(L_{1i},L_{2i},L_{3i})|J(L_{1i},L_{2i},L_{3i})| \qquad (1.146)$$

where data on the weighting coefficients W_i and area coordinates L_{1i}, L_{2i}, L_{3i} are presented in Table 1.9.

Table 1.9: Numerical integration formulas for triangles.

n	i	W_i	(L_{1i}, L_{2i}, L_{3i})	
1	1	1	(α,α,α)	$\alpha = 1/3$
3	1	1/3	$(\alpha,\alpha,0)$	
	2	1/3	$(0,\alpha,\alpha)$	$\alpha = 1/2$
	3	1/3	$(\alpha,0,\alpha)$	
4	1	$-27/48$	(α,α,α)	$\alpha = 1/3$
	2	25/48	(β,γ,γ)	$\beta = 0.6$
	3	25/48	(γ,β,γ)	$\gamma = 0.2$
	4	25/48	(γ,γ,β)	
7	1	0.225	(α,α,α)	$\alpha = 1/3$
	2	0.13239415	(β,γ,γ)	$\beta = 0.05971587$
	3	0.13239415	(γ,β,γ)	$\gamma = 0.47014206$
	4	0.13239415	(γ,γ,β)	$\delta = 0.79742669$
	5	0.12593918	$(\delta,\varepsilon,\varepsilon)$	$\varepsilon = 0.10128651$
	6	0.12593918	$(\varepsilon,\delta,\varepsilon)$	
	7	0.12593918	$(\varepsilon,\varepsilon,\delta)$	

1.5.3   Three-dimensional integration

Using the local coordinates ξ, η, ζ, three-dimensional integration may be expressed as

$$\iiint f(x,y,z)\,dx\,dy\,dz = \iiint f(\xi,\eta,\zeta)|J(\xi,\eta,\zeta)|d\xi\,d\eta\,d\zeta\,. \qquad (1.147)$$

For a tetrahedron, in terms of the volume coordinates (see Eq. (1.107)) the numerical integration formula derived by Hammer *et al.* is of the form

$$\iiint_e f(L_1, L_2, L_3, L_4)|J(L_1, L_2, L_3, L_4)|d\xi\,d\eta\,d\zeta$$

$$= \sum_{i=1}^{n} \frac{1}{6} W_i f(L_{1i}, L_{2i}, L_{3i}, L_{4i})|J(L_{1i}, L_{2i}, L_{3i}, L_{4i})| \qquad (1.148)$$

where data on the weighting coefficients W_i and volume coordinates L_{1i}, L_{2i}, L_{3i}, L_{4i} are presented in Table 1.10.

Table 1.10: Numerical integration formulas for tetrahedra.

n	i	W_i	$(L_{1i}, L_{2i}, L_{3i}, L_{4i})$	
1	1	1	$(\alpha, \alpha, \alpha, \alpha)$	$\alpha = 1/4$
	1	1/4	$(\alpha, \beta, \beta, \beta)$	
4	2	1/4	$(\beta, \alpha, \beta, \beta)$	$\alpha = 0.58541020$
	3	1/4	$(\beta, \beta, \alpha, \beta)$	$\beta = 0.13819660$
	4	1/4	$(\beta, \beta, \beta, \alpha)$	
	1	$-4/5$	$(\alpha, \alpha, \alpha, \alpha)$	
	2	9/20	$(\beta, \gamma, \gamma, \gamma)$	$\alpha = 1/4$
5	3	9/20	$(\gamma, \beta, \gamma, \gamma)$	$\beta = 1/3$
	4	9/20	$(\gamma, \gamma, \beta, \gamma)$	$\gamma = 1/6$
	5	9/20	$(\gamma, \gamma, \gamma, \beta)$	

1.6   Special Elements

Current applications of the FEM show a trend towards the enhancement of solution accuracy through the utilization of elements with special interpolation functions. Many different special-purpose elements have been developed. A survey of this type of special element is now presented. This class includes isoparametric elements, edge elements, infinite elements, and boundary elements.

In what follows we shall discuss two-dimensional special elements which have found a great deal of practical application.

1.6.1   Isoparametric elements

Triangular elements are widely used in finite element analysis because any polygonal object can be decomposed into a set of triangles without approximation, but they do not lend themselves well to the modeling of curved shapes. The use of so-called isoparametric elments, which have curved sides, can often alleviate the problems encountered in geometric modeling, by shaping the elements to fit the real geometry.

In the isoparametric element, the polynomial order representing the geometry of an element is coincident with that interpolating the field in an element. In the isoparametric quadratic element as shown in Fig. 1.7, the field ϕ in an element and the coordinate on a curvilinear side can be expressed as

$$\phi = \sum_{i=1}^{6} N_i \phi_i = \{N\}^{\mathrm{T}} \{\phi\}_e \qquad (1.149)$$

$$x = \sum_{i=1}^{6} N_i x_i = \{N\}^{\mathrm{T}} \{x\}_e \qquad (1.150)$$

$$y = \sum_{i=1}^{6} N_i y_i = \{N\}^{\mathrm{T}} \{y\}_e \qquad (1.151)$$

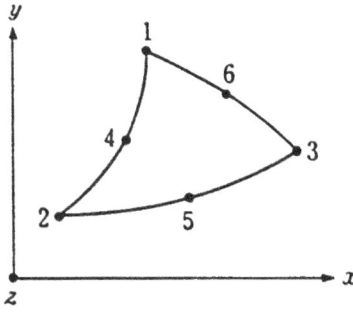

Figure 1.7: Isoparametric quadratic element.

with

$$\{N\} = \begin{bmatrix} L_1(2L_1 - 1) \\ L_2(2L_2 - 1) \\ L_3(2L_3 - 1) \\ 4L_1L_2 \\ 4L_2L_3 \\ 4L_3L_1 \end{bmatrix} \tag{1.152}$$

where x_i, y_i are the Cartesian coordinates of the node i $(i = 1, 2, \cdots, 6)$ of the element and L_1, L_2, L_3 are the area coordinates (see Table 1.2).

Defining the local coordinates ξ, η (see Eq. (1.80)), the transformation relation for differentiation is given by

$$\begin{bmatrix} \partial/\partial\xi \\ \partial/\partial\eta \end{bmatrix} = [J] \begin{bmatrix} \partial/\partial x \\ \partial/\partial y \end{bmatrix} = \begin{bmatrix} J_{11} & J_{12} \\ J_{21} & J_{22} \end{bmatrix} \begin{bmatrix} \partial/\partial x \\ \partial/\partial y \end{bmatrix} \tag{1.153}$$

or

$$\begin{bmatrix} \partial/\partial x \\ \partial/\partial y \end{bmatrix} = [J]^{-1} \begin{bmatrix} \partial/\partial\xi \\ \partial/\partial\eta \end{bmatrix} = \frac{1}{|J|} \begin{bmatrix} J_{22} & -J_{12} \\ -J_{21} & J_{11} \end{bmatrix} \begin{bmatrix} \partial/\partial\xi \\ \partial/\partial\eta \end{bmatrix} \tag{1.154}$$

with

$$J_{11} = \frac{\partial x}{\partial \xi} = \frac{\partial \{N\}^{\mathrm{T}}}{\partial \xi} \{x\}_e \tag{1.155a}$$

$$J_{12} = \frac{\partial y}{\partial \xi} = \frac{\partial \{N\}^{\mathrm{T}}}{\partial \xi} \{y\}_e \tag{1.155b}$$

$$J_{21} = \frac{\partial x}{\partial \eta} = \frac{\partial \{N\}^{\mathrm{T}}}{\partial \eta} \{x\}_e \tag{1.155c}$$

$$J_{22} = \frac{\partial y}{\partial \eta} = \frac{\partial \{N\}^{\mathrm{T}}}{\partial \eta} \{y\}_e \tag{1.155d}$$

$$|J| = J_{11}J_{22} - J_{12}J_{21} . \tag{1.156}$$

The transformation relation for integration is given by

$$\iiint_e f(x, y) \, dx \, dy = \int_0^1 \int_0^{1-\xi} f(\xi, \eta) |J| \, d\xi \, d\eta . \tag{1.157}$$

From the above relations the integrals necessary to construct element matrices are reduced to the form

$$\iint_e \{N\}\{N\}^{\mathrm{T}} \, dx \, dy = \iint_e \{N\}\{N\}^{\mathrm{T}} |J| \, d\xi \, d\eta \tag{1.158}$$

$$\iint_e \{N_x\}\{N_x\}^T \, dx \, dy \quad = \quad \iint_e (J_{22}\{N_\xi\} - J_{12}\{N_\eta\})$$
$$\cdot (J_{22}\{N_\xi\}^T - J_{12}\{N_\eta\}^T)|J|^{-1} \, d\xi \, d\eta$$

$$(1.159)$$

$$\iint_e \{N_y\}\{N_y\}^T \, dx \, dy \quad = \quad \iint_e (-J_{21}\{N_\xi\} + J_{11}\{N_\eta\})$$
$$\cdot (-J_{21}\{N_\xi\}^T + J_{11}\{N_\eta\}^T)|J|^{-1} \, d\xi \, d\eta$$

$$(1.160)$$

$$\iint_e \{N_x\}\{N\}^T \, dx \, dy \quad = \quad \iint_e (J_{22}\{N_\xi\} - J_{12}\{N_\eta\})\{N\}^T \, d\xi \, d\eta$$

$$(1.161)$$

$$\iint_e \{N_y\}\{N\}^T \, dx \, dy \quad = \quad \iint_e (-J_{21}\{N_\xi\} + J_{11}\{N_\eta\})\{N\}^T \, d\xi \, d\eta$$

$$(1.162)$$

with

$$\{N_\xi\} \quad = \quad \frac{\partial\{N\}}{\partial \xi} = \frac{\partial\{N\}}{\partial L_1} = \begin{bmatrix} 4L_1 - 1 \\ 0 \\ 1 - 4L_3 \\ 4L_2 \\ -4L_2 \\ 4(L_3 - L_1) \end{bmatrix} \qquad (1.163)$$

$$\{N_\eta\} \quad = \quad \frac{\partial\{N\}}{\partial \eta} = \frac{\partial\{N\}}{\partial L_2} = \begin{bmatrix} 0 \\ 4L_2 - 1 \\ 1 - 4L_3 \\ 4L_1 \\ 4(L_3 - L_2) \\ -4L_1 \end{bmatrix}. \qquad (1.164)$$

These integrals, Eqs. (1.158) to (1.162), can be straightforwardly evaluated by utilizing the numerical integration formula, Eq. (1.146).

1.6.2 Edge elements

The electromagnetic fields have to be tangentially continuous across material interfaces.

In the edge element, as shown in Fig. 1.8, the tangential continuity can be straightforwardly imposed. The six nodes described in the element consist of the three corner and three side points. The corner points 1 to 3 are for the longitudinal component ϕ_z (E_z or H_z), while the side points

4 to 6 are for the tangential component ϕ_t (E_t or H_t). The longitudinal component ϕ_z is approximated by a complete polynomial of first order:

$$\phi_z = \{N(x,y)\}^{\mathrm{T}}\{\phi_z\}_e = \{N\}^{\mathrm{T}}\{\phi_z\}_e \qquad (1.165)$$

where $\{N\}$ is the ordinary shape function vector for the linear triangular element (see Fig. 1.3(a) and Table 1.2). The transverse components ϕ_x (E_x or H_x) and ϕ_y (E_y or H_y) are approximated by a linear function of y and x, respectively:

$$\begin{aligned}
\phi_x &= \{U(y)\}^{\mathrm{T}}\{\phi_t\}_e = \{U\}^{\mathrm{T}}\{\phi_t\}_e & (1.166) \\
\phi_y &= \{V(x)\}^{\mathrm{T}}\{\phi_t\}_e = \{V\}^{\mathrm{T}}\{\phi_t\}_e & (1.167)
\end{aligned}$$

with

$$\{U\} = \begin{bmatrix} \tilde{a}_1 + \tilde{c}_1 y \\ \tilde{a}_2 + \tilde{c}_2 y \\ \tilde{a}_3 + \tilde{c}_3 y \end{bmatrix} \qquad (1.168)$$

$$\{V\} = \begin{bmatrix} \tilde{b}_1 - \tilde{c}_1 x \\ \tilde{b}_2 - \tilde{c}_2 x \\ \tilde{b}_3 - \tilde{c}_3 x \end{bmatrix} \qquad (1.169)$$

where the coefficients $\tilde{a}_k$, $\tilde{b}_k$, $\tilde{c}_k$ are given by

$$\begin{aligned}
\tilde{a}_k &= [(y_{m+3}\cos\theta_{m+3} - x_{m+3}\sin\theta_{m+3})\sin\theta_{l+3} \\
&\quad - (y_{l+3}\cos\theta_{l+3} - x_{l+3}\sin\theta_{l+3})\sin\theta_{m+3}]/\Delta & (1.170) \\
\tilde{b}_k &= [(y_{l+3}\cos\theta_{l+3} - x_{l+3}\sin\theta_{l+3})\cos\theta_{m+3} \\
&\quad - (y_{m+3}\cos\theta_{m+3} - x_{m+3}\sin\theta_{m+3})\cos\theta_{l+3}]/\Delta & (1.171) \\
\tilde{c}_k &= (\cos\theta_{l+3}\sin\theta_{m+3} - \cos\theta_{m+3}\sin\theta_{l+3})/\Delta & (1.172)
\end{aligned}$$

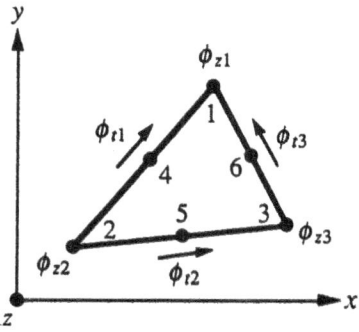

Figure 1.8: Edge element.

with

$$0 \leq \theta_{k+3} = \tan^{-1}\{(y_k - y_l)/(x_k - x_l)\} < \pi \qquad (1.173)$$

$$\Delta = \sum_{k=1}^{3}(y_{k+3}\cos\theta_{k+3} - x_{k+3}\sin\theta_{k+3})$$
$$\cdot(\cos\theta_{l+3}\sin\theta_{m+3} - \cos\theta_{m+3}\sin\theta_{l+3}) . \qquad (1.174)$$

Here the subscripts k, l, m always progress modulo 3, i.e., cyclically around the three vertices of the triangle, and x_i, y_i are the Cartesian coordinates of the node i ($i = 1, 2, \cdots, 6$) of the element.

From Eqs. (1.74), (1.86), and (1.87) the integrals necessary to construct element matrices are calculated as follows:

$$\left[\iint_e \{U\}\{U\}^{\mathrm{T}}\, dx\, dy\right]_{kl} = A_e \tilde{a}_k \tilde{a}_l + A_e y_c(\tilde{a}_k \tilde{c}_l + \tilde{c}_k \tilde{a}_l)$$
$$+ \frac{1}{12} A_e \tilde{c}_k \tilde{c}_l (y_1^2 + y_2^2 + y_3^2 + 9y_c^2)$$
$$(1.175)$$

$$\left[\iint_e \{V\}\{V\}^{\mathrm{T}}\, dx\, dy\right]_{kl} = A_e \tilde{b}_k \tilde{b}_l - A_e x_c(\tilde{b}_k \tilde{c}_l + \tilde{c}_k \tilde{b}_l)$$
$$+ \frac{1}{12} A_e \tilde{c}_k \tilde{c}_l (x_1^2 + x_2^2 + x_3^2 + 9x_c^2)$$
$$(1.176)$$

$$\left[\iint_e \{U_y\}\{U_y\}^{\mathrm{T}}\, dx\, dy\right]_{kl} = \left[\iint_e \{V_x\}\{V_x\}^{\mathrm{T}}\, dx\, dy\right]_{kl}$$
$$= -\left[\iint_e \{U_y\}\{V_x\}^{\mathrm{T}}\, dx\, dy\right]_{kl}$$
$$= -\left[\iint_e \{V_x\}\{U_y\}^{\mathrm{T}}\, dx\, dy\right]_{kl}$$
$$= A_e \tilde{c}_k \tilde{c}_l \qquad (1.177)$$

$$\left[\iint_e \{N_x\}\{U\}^{\mathrm{T}}\, dx\, dy\right]_{kl} = \frac{1}{2}b_k(\tilde{a}_l + \tilde{c}_l y_c) \qquad (1.178)$$

$$\left[\iint_e \{N_y\}\{V\}^{\mathrm{T}}\, dx\, dy\right]_{kl} = \frac{1}{2}c_k(\tilde{b}_l - \tilde{c}_l x_c) \qquad (1.179)$$

with

$$x_c = \frac{1}{3}(x_1 + x_2 + x_3) \qquad (1.180)$$

$$y_c = \frac{1}{3}(y_1 + y_2 + y_3) \qquad (1.181)$$

where $[\cdot]_{kl}$ $(kl = 11, 12, \cdots, 33)$ indicates the (k, l) component of the matrix $[\cdot]$, b_k and c_k are given in Eqs. (1.78) and (1.79), respectively, and $\{N_x\} \equiv \partial\{N\}/\partial x$, $\{N_y\} \equiv \partial\{N\}/\partial y$, $\{U_y\} \equiv d\{U\}/dy$, and $\{V_x\} \equiv d\{V\}/dx$. Note that the tangential component, $\phi_t = \phi_x \cos\theta + \phi_y \sin\theta$, is continuous along the interelement boundaries and is constant on each side of triangles.

1.6.3   Infinite elements

It is well recognized that difficulty is frequently encountered when one wants to solve unbounded field problems using finite elements. To overcome this difficulty, these unbounded domains have in the past been dealt with in various ways. To date the main methods in optical waveguide problems have been simple truncation, the use of analytical far-field solutions, the infinite element approach, the exterior finite element approach, and the conformal mapping technique. The simplest of these techniques is undoubtedly the simple truncation, in which the unbounded structure is replaced by a corresponding bounded one by setting artificial boundaries in the position far from the guiding region. However, this technique involves a very large number of nodes when the field extends farther away from the guiding region. Among other methods the infinite element approach, in which the bounded region and the semi-infinite homogeneous region are divided into standard finite elements and infinite elements, respectively, is often simple and economical and has now been applied successfully to a wide range of problems.

Infinite elements are obtained by moving one side of an element to infinity. In this process, however, an integrand for computing element matrices diverges. Thus, the shape functions for such an element should realistically represent the fields and should be square integrable over an infinite element area. To satisfy the radiation condition, decay-type infinite elements are generally used.

Consider strip-like infinite elements shown in Figs. 1.9 to 1.11 and expand the field ϕ in each element as

$$\{\phi\} = \{N\}^{\mathrm{T}}\{\phi\}_e \,. \tag{1.182}$$

The shape function vector $\{N\}$ is written as follows:

(1) Infinite elements of type i:

$$\{N\} = f(x)f(y) \tag{1.183}$$

with

$$f(x) = \exp\{-\alpha_x(x - x_0)\} \tag{1.184}$$
$$f(y) = \exp\{-\alpha_y(y - y_0)\} \tag{1.185}$$

where α_x and α_y are called the decay parameters.

(2) Linear infinite elements of type ii:

$$\{N\} = \begin{bmatrix} L_1 f(x) \\ L_2 f(x) \end{bmatrix} \qquad (1.186)$$

with

$$\begin{bmatrix} 1 \\ y \end{bmatrix} = \begin{bmatrix} 1 & 1 \\ y_1 & y_2 \end{bmatrix} \begin{bmatrix} L_1 \\ L_2 \end{bmatrix} \qquad (1.187)$$

where y_k is the Cartesian coordinate of the vertex k $(k = 1, 2)$ of the element.

(3) Quadratic infinite elements of type ii:

$$\{N\} = \begin{bmatrix} L_1(2L_1 - 1)f(x) \\ L_2(2L_2 - 1)f(x) \\ 4L_1 L_2 f(x) \end{bmatrix} . \qquad (1.188)$$

(4) Linear infinite elements of type iii:

$$\{N\} = \begin{bmatrix} L_1 f(y) \\ L_2 f(y) \end{bmatrix} \qquad (1.189)$$

with

$$\begin{bmatrix} 1 \\ x \end{bmatrix} = \begin{bmatrix} 1 & 1 \\ x_1 & x_2 \end{bmatrix} \begin{bmatrix} L_1 \\ L_2 \end{bmatrix} \qquad (1.190)$$

where x_k is the Cartesian coordinate of the vertex k $(k = 1, 2)$ of the element.

(5) Quadratic infinite elements of type iii:

$$\{N\} = \begin{bmatrix} L_1(2L_1 - 1)f(y) \\ L_2(2L_2 - 1)f(y) \\ 4L_1 L_2 f(y) \end{bmatrix} . \qquad (1.191)$$

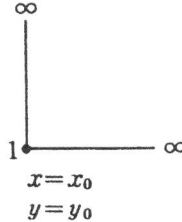

Figure 1.9: Infinite element (type i).

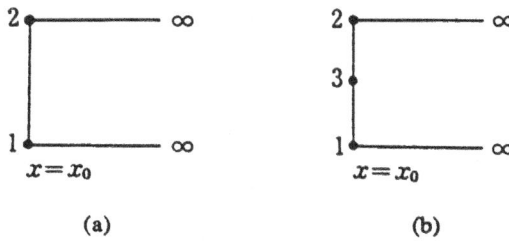

(a)   (b)

Figure 1.10: Infinite elements (type ii). (a) Linear element. (b) Quadratic element.

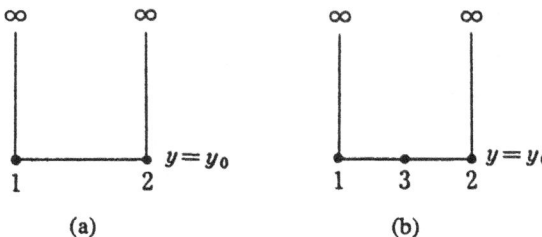

(a)   (b)

Figure 1.11: Infinite elements (type iii). (a) Linear element. (b) Quadratic element.

The integrals necessary to construct element matrices are calculated as follows:

(1) Infinite elements of type i:

$$\iint_e \{N\}\{N\}^\mathrm{T}\, dx\, dy \;=\; \frac{1}{4\alpha_x\alpha_y} \tag{1.192}$$

$$\iint_e \{N_x\}\{N_x\}^\mathrm{T}\, dx\, dy \;=\; \frac{\alpha_x}{4\alpha_y} \tag{1.193}$$

$$\iint_e \{N_y\}\{N_y\}^\mathrm{T}\, dx\, dy \;=\; \frac{\alpha_y}{4\alpha_x} \tag{1.194}$$

$$\iint_e \{N_x\}\{N_y\}^\mathrm{T}\, dx\, dy \;=\; \frac{1}{4} \tag{1.195}$$

$$\iint_e \{N_x\}\{N\}^\mathrm{T}\, dx\, dy \;=\; -\frac{1}{4\alpha_y} \tag{1.196}$$

$$\iint_e \{N_y\}\{N\}^\mathrm{T}\, dx\, dy \;=\; -\frac{1}{4\alpha_x}. \tag{1.197}$$

(2) Linear infinite elements of type ii:

$$\iint_e \{N\}\{N\}^\mathrm{T}\, dx\, dy \;=\; \frac{l_e}{12\alpha_x} \begin{bmatrix} 2 & 1 \\ 1 & 2 \end{bmatrix} \tag{1.198}$$

$$\iint_e \{N_x\}\{N_x\}^T \, dx \, dy = \frac{\alpha_x l_e}{12} \begin{bmatrix} 2 & 1 \\ 1 & 2 \end{bmatrix} \qquad (1.199)$$

$$\iint_e \{N_y\}\{N_y\}^T \, dx \, dy = \frac{1}{2\alpha_x l_e} \begin{bmatrix} 1 & -1 \\ -1 & 1 \end{bmatrix} \qquad (1.200)$$

$$\iint_e \{N_x\}\{N_y\}^T \, dx \, dy = -\frac{1}{4} \begin{bmatrix} -1 & 1 \\ -1 & 1 \end{bmatrix} \qquad (1.201)$$

$$\iint_e \{N_x\}\{N\}^T \, dx \, dy = -\frac{l_e}{12} \begin{bmatrix} 2 & 1 \\ 1 & 2 \end{bmatrix} \qquad (1.202)$$

$$\iint_e \{N_y\}\{N\}^T \, dx \, dy = \frac{1}{4\alpha_x} \begin{bmatrix} -1 & -1 \\ 1 & 1 \end{bmatrix} \qquad (1.203)$$

with

$$l_e = y_2 - y_1 . \qquad (1.204)$$

(3) Quadratic infinite elements of type ii:

$$\iint_e \{N\}\{N\}^T \, dx \, dy = \frac{l_e}{60\alpha_x} \begin{bmatrix} 4 & -1 & 2 \\ -1 & 4 & 2 \\ 2 & 2 & 16 \end{bmatrix} \qquad (1.205)$$

$$\iint_e \{N_x\}\{N_x\}^T \, dx \, dy = \frac{\alpha_x l_e}{60} \begin{bmatrix} 4 & -1 & 2 \\ -1 & 4 & 2 \\ 2 & 2 & 16 \end{bmatrix} \qquad (1.206)$$

$$\iint_e \{N_y\}\{N_y\}^T \, dx \, dy = \frac{1}{6\alpha_x l_e} \begin{bmatrix} 7 & 1 & -8 \\ 1 & 7 & -8 \\ -8 & -8 & 16 \end{bmatrix} \qquad (1.207)$$

$$\iint_e \{N_x\}\{N_y\}^T \, dx \, dy = -\frac{1}{12} \begin{bmatrix} -3 & -1 & 4 \\ 1 & 3 & -4 \\ -4 & 4 & 0 \end{bmatrix} \qquad (1.208)$$

$$\iint_e \{N_x\}\{N\}^T \, dx \, dy = -\frac{l_e}{60} \begin{bmatrix} 4 & -1 & 2 \\ -1 & 4 & 2 \\ 2 & 2 & 16 \end{bmatrix} \qquad (1.209)$$

$$\iint_e \{N_y\}\{N\}^T \, dx \, dy = \frac{1}{12\alpha_x} \begin{bmatrix} -3 & 1 & -4 \\ -1 & 3 & 4 \\ 4 & -4 & 0 \end{bmatrix} . \qquad (1.210)$$

(4) Linear infinite elements of type iii:

$$\iint_e \{N\}\{N\}^T \, dx \, dy = \frac{l_e}{12\alpha_y} \begin{bmatrix} 2 & 1 \\ 1 & 2 \end{bmatrix} \qquad (1.211)$$

$$\iint_e \{N_x\}\{N_x\}^T dx\, dy \;=\; \frac{1}{2\alpha_y l_e}\begin{bmatrix} 1 & -1 \\ -1 & 1 \end{bmatrix} \qquad (1.212)$$

$$\iint_e \{N_y\}\{N_y\}^T dx\, dy \;=\; \frac{\alpha_y l_e}{12}\begin{bmatrix} 2 & 1 \\ 1 & 2 \end{bmatrix} \qquad (1.213)$$

$$\iint_e \{N_x\}\{N_y\}^T dx\, dy \;=\; -\frac{1}{4}\begin{bmatrix} -1 & -1 \\ 1 & 1 \end{bmatrix} \qquad (1.214)$$

$$\iint_e \{N_x\}\{N\}^T dx\, dy \;=\; \frac{1}{4\alpha_y}\begin{bmatrix} -1 & -1 \\ 1 & 1 \end{bmatrix} \qquad (1.215)$$

$$\iint_e \{N_y\}\{N\}^T dx\, dy \;=\; -\frac{l_e}{12}\begin{bmatrix} 2 & 1 \\ 1 & 2 \end{bmatrix} \qquad (1.216)$$

with

$$l_e = x_2 - x_1 \,. \qquad (1.217)$$

(5) Quadratic infinite elements of type iii:

$$\iint_e \{N\}\{N\}^T dx\, dy \;=\; \frac{l_e}{60\alpha_y}\begin{bmatrix} 4 & -1 & 2 \\ -1 & 4 & 2 \\ 2 & 2 & 16 \end{bmatrix} \qquad (1.218)$$

$$\iint_e \{N_x\}\{N_x\}^T dx\, dy \;=\; \frac{1}{60\alpha_y l_e}\begin{bmatrix} 7 & 1 & -8 \\ 1 & 7 & -8 \\ -8 & -8 & 16 \end{bmatrix} \qquad (1.219)$$

$$\iint_e \{N_y\}\{N_y\}^T dx\, dy \;=\; \frac{\alpha_y l_e}{60}\begin{bmatrix} 4 & -1 & 2 \\ -1 & 4 & 2 \\ 2 & 2 & 16 \end{bmatrix} \qquad (1.220)$$

$$\iint_e \{N_x\}\{N_y\}^T dx\, dy \;=\; -\frac{1}{12}\begin{bmatrix} -3 & 1 & -4 \\ -1 & 3 & 4 \\ 4 & -4 & 0 \end{bmatrix} \qquad (1.221)$$

$$\iint_e \{N_x\}\{N\}^T dx\, dy \;=\; \frac{1}{12\alpha_y}\begin{bmatrix} -3 & 1 & -4 \\ -1 & 3 & 4 \\ 4 & -4 & 0 \end{bmatrix} \qquad (1.222)$$

$$\iint_e \{N_y\}\{N\}^T dx\, dy \;=\; -\frac{l_e}{60}\begin{bmatrix} 4 & -1 & 2 \\ -1 & 4 & 2 \\ 2 & 2 & 16 \end{bmatrix} . \qquad (1.223)$$

The decay-type infinite element approach can reduce the finite element domain compared with the artificial boundary approach, but it presents difficulties in choosing the decay parameters. The best value of decay parameter depends on the operating frequency and the kind of mode.

1.6.4   Boundary elements

The boundary element method (BEM) is one of the boundary-type methods based on the integral equation method and, therefore, if the region to be analyzed is homogeneous, then it requires nodes, necessary for calculation, on its boundary only.[6] So the problem can be treated with one less dimension. Moreover, it can handle unbounded field problems easily, so that it is suitable for the optical waveguide analysis which often includes unbounded regions.

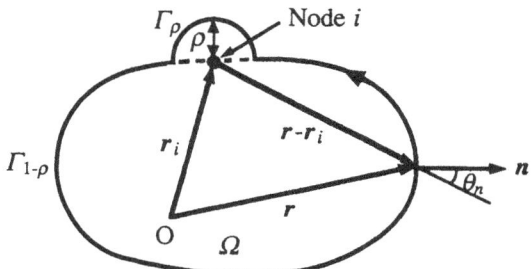

Figure 1.12: Two-dimensional region Ω surrounded by boundary Γ.

Consider the region Ω with the boundary Γ shown in Fig. 1.12 and the following two-dimensional Helmholtz equation:

$$\frac{\partial^2 \phi}{\partial x^2} + \frac{\partial^2 \phi}{\partial y^2} + k^2 \phi = 0 . \tag{1.224}$$

Using the fundamental solution (Green's fuction) G_i and Green's formula, from Eq. (1.224) we obtain

$$\phi_i + \int_\Gamma \frac{\partial G_i}{\partial n} \phi \, d\Gamma = \int_\Gamma G_i \frac{\partial \phi}{\partial n} \, d\Gamma \tag{1.225}$$

with

$$G_i = \frac{1}{4j} H_0^{(2)}(k|\mathbf{r} - \mathbf{r}_i|) \tag{1.226}$$

$$\frac{\partial G_i}{\partial n} = -\frac{k}{4j} H_1^{(2)}(k|\mathbf{r} - \mathbf{r}_i|) \cos\theta_n \tag{1.227}$$

where ϕ_i is the value of ϕ at node i, $\partial/\partial n$ is the outward normal derivative, $H_0^{(2)}$ and $H_1^{(2)}$ are the zeroth- and first-order Hankel functions of the second kind, respectively, θ_n is the angle between the vector $\mathbf{r} - \mathbf{r}_i$ and the outward unit normal vector $\mathbf{n}$.

Remembering that the node i is placed on the boundary Γ and considering that the integration path Γ is replaced by another path $\Gamma_\rho + \Gamma_{1-\rho}$ that goes around the node i as shown in Fig. 1.12, we estimate the integration over the path Γ_ρ as follows:

$$
\begin{aligned}
\lim_{\rho \to 0} \int \frac{\partial G_i}{\partial n} \phi \, d\Gamma &= \lim_{\rho \to 0} \left\{ \rho(2\pi - \theta_i)\phi \left[-\frac{k}{4j} H_1^{(2)}(k\rho) \right] \right\} \\
&= j\frac{k(2\pi - \theta_i)}{4} \lim_{\rho \to 0} \left\{ \rho\phi \left[\frac{k\rho}{2} - j\left(-\frac{1}{\pi}\frac{2}{k\rho} \right) \right] \right\} \\
&= \left(\frac{\theta_i}{2\pi} - 1 \right) \phi_i \qquad\qquad\qquad (1.228)
\end{aligned}
$$

$$
\begin{aligned}
\lim_{\rho \to 0} \int \frac{\partial \phi}{\partial n} G_i \, d\Gamma &= \lim_{\rho \to 0} \left\{ \rho(2\pi - \theta_i)\frac{\partial \phi}{\partial n} \left[\frac{1}{4j} H_0^{(2)}(k\rho) \right] \right\} \\
&= -j\frac{2\pi - \theta_i}{4} \lim_{\rho \to 0} \left\{ \rho\frac{\partial \phi}{\partial n} \left[1 - j\frac{2}{\pi}\left(\ln\frac{k\rho}{2} + \gamma \right) \right] \right\} \\
&= 0 \qquad\qquad\qquad\qquad\qquad\qquad\qquad (1.229)
\end{aligned}
$$

where θ_i is the internal angle of the boundary at the node i, and γ is the Euler number and is given by

$$
\gamma = 0.577215 \,. \qquad\qquad\qquad (1.230)
$$

From Eqs. (1.228) and (1.229) we obtain for Eq. (1.225)

$$
\frac{\theta_i}{2\pi}\phi_i + \fint_\Gamma \frac{\partial G_i}{\partial n} \phi \, d\Gamma = \fint_\Gamma \frac{\partial \phi}{\partial n} G_i \, d\Gamma \qquad\qquad (1.231)
$$

with

$$
\fint_\Gamma d\Gamma = \lim_{\rho \to 0} \int_{\Gamma_{1-\rho}} d\Gamma \qquad\qquad (1.232)
$$

where $\fint$ denotes Cauchy's principal value of integration.

Dividing the boundary Γ into boundary elements as shown in Fig. 1.13, we expand ϕ and $\partial\phi/\partial n \equiv \psi$ within each element in terms of ϕ_i and ψ_i at the node i, respectively, namely

$$
\begin{aligned}
\phi &= \{N\}^{\mathrm{T}}\{\phi\}_e \qquad\qquad\qquad (1.233) \\
\partial\phi/\partial n &= \psi = \{N\}^{\mathrm{T}}\{\psi\}_e. \qquad\qquad (1.234)
\end{aligned}
$$

The shape function vector $\{N\}$ is written as follows:

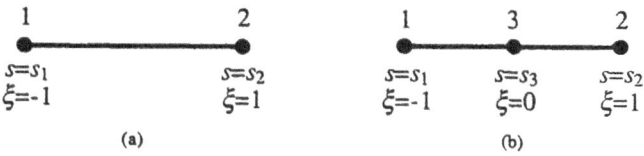

Figure 1.13: Boundary elements. (a) Linear element. (b) Quadratic element.

(1) Linear elements:

$$\{N\} = \begin{bmatrix} a_1 + b_1\xi \\ a_2 + b_2\xi \end{bmatrix} = \begin{bmatrix} (1-\xi)/2 \\ (1+\xi)/2 \end{bmatrix} \tag{1.235}$$

where the relation equation between the local coordinate ξ and the global coordinate s along the boundary Γ is given by

$$s = \frac{1}{2}(s_2 - s_1)\xi + \frac{1}{2}(s_2 + s_1). \tag{1.236}$$

Here s_k is the global coordinate of the edge k $(k = 1, 2)$ of the element.

(2) Quadratic elements:

$$\{N\} = \begin{bmatrix} a_1 + b_1\xi + c_1\xi^2 \\ a_2 + b_2\xi + c_2\xi^2 \\ a_3 + b_3\xi + c_3\xi^2 \end{bmatrix} = \begin{bmatrix} -\xi(1-\xi)/2 \\ \xi(1+\xi)/2 \\ 1 - \xi^2 \end{bmatrix}. \tag{1.237}$$

Substituting Eqs. (1.233) and (1.234) into Eq. (1.231), we obtain

$$\frac{\theta_i}{2\pi}\phi_i + \sum_e \{h\}_e^T\{\phi\}_e = \sum_e \{g\}_e^T\{\psi\}_e \tag{1.238}$$

where $\sum_e$ extends over all different elements. When the node i does not belong to the eth element, h_j and g_j are calculated with Gaussian integration as

$$h_j = \frac{l_e}{2} \int_{-1}^{1} -\frac{k}{4j} H_1^{(2)}(k|\mathbf{r} - \mathbf{r}_i|) N_j \cos\theta_n \, d\xi \tag{1.239}$$

$$g_j = \frac{l_e}{2} \int_{-1}^{1} \frac{1}{4j} H_0^{(2)}(k|\mathbf{r} - \mathbf{r}_i|) N_j \, d\xi \tag{1.240}$$

where l_e is the length of the element. When the node i belongs to the eth element, calculations of h_j and g_j involve the limitation of $\rho \to 0$. In this case, $\cos\theta_n = 0$, so that

$$h_j = 0. \tag{1.241}$$

For the case 1, 2, or 3 where the node i coincides with the node 1, 2, or 3 of the eth element, respectively, g_j is calculated as follows:

(1) Linear elements:

$$
\begin{aligned}
g_j &= \lim_{\rho \to 0} \int_{s_1+\rho}^{s_2} \frac{N_j}{4j} H_0^{(2)}(k|s - s_1|)\, ds \\
&= \lim_{\rho \to 0} \frac{l_e}{2} \int_{-1+\rho}^{1} \frac{N_j}{4j} H_0^{(2)}\left(\frac{kl_e}{2}|\xi + 1|\right) d\xi \\
&= \lim_{\rho \to 0} \frac{l_e}{2} \int_{\rho}^{2} [a_j + b_j(\eta - 1)] \frac{1}{4j} H_0^{(2)}\left(\frac{kl_e}{2}\eta\right) d\eta \\
&= (l_e/2)\{(a_j - b_j)I_0(2) \\
&\quad + b_j[I_1(2) - 2/(\pi k^2 l_e^2)]\} \qquad \text{for case 1} \qquad (1.242\text{a}) \\
g_j &= \lim_{\rho \to 0} \int_{s_1}^{s_2-\rho} \frac{N_j}{4j} H_0^{(2)}(k|s - s_2|)\, ds \\
&= \lim_{\rho \to 0} \frac{l_e}{2} \int_{-1}^{1-\rho} \frac{N_j}{4j} H_0^{(2)}\left(\frac{kl_e}{2}|\xi - 1|\right) d\xi \\
&= \lim_{\rho \to 0} \frac{l_e}{2} \int_{\rho}^{2} [a_j + b_j(1 - \eta)] \frac{1}{4j} H_0^{(2)}\left(\frac{kl_e}{2}\eta\right) d\eta \\
&= (l_e/2)\{(a_j + b_j)I_0(2) \\
&\quad - b_j[I_1(2) - 2/(\pi k^2 l_e^2)]\} \qquad \text{for case 2} \qquad (1.242\text{b})
\end{aligned}
$$

where the Hankel function is expanded into infinite series and is integrated term-by-term

$$
\begin{aligned}
I_0(\eta) &= \int \frac{1}{4j} H_0^{(2)}\left(\frac{kl_e}{2}\eta\right) d\eta \\
&= -\frac{\eta}{4} \sum_{\mu=0}^{\infty} \frac{(-1)^\mu}{(2\mu + 1)(\mu!)^2} \left(\frac{kl_e}{4}\eta\right)^{2\mu} \\
&\quad \cdot \left\{\frac{2}{\pi}\left[\gamma + \ln\left(\frac{kl_e}{4}\eta\right) - \frac{1}{2\mu + 1} - \sum_{\nu=1}^{\mu} \frac{1}{\nu}\right] + j\right\}
\end{aligned}
$$

$$
(1.243)
$$

$$
\begin{aligned}
I_1(\eta) &= \int \frac{\eta}{4j} H_0^{(2)}\left(\frac{kl_e}{2}\eta\right) d\eta \\
&= \frac{1}{4j} \frac{2\eta}{kl_e} H_1^{(2)}\left(\frac{kl_e}{2}\eta\right)
\end{aligned}
$$

$$
(1.244)
$$

(2) Quadratic elements:

$$
\begin{aligned}
g_j &= \lim_{\rho \to 0} \frac{l_e}{2} \int_{\rho}^{2} [a_j + b_j(\eta - 1) + c_j(\eta - 1)^2] \frac{1}{4j} H_0^{(2)}\left(\frac{kl_e}{2}\eta\right) d\eta \\
&= (l_e/2)\{(a_j - b_j + c_j)I_0(2) \\
&\quad + (b_j - 2c_j)[I_1(2) - 2/(\pi k^2 l_e^2)] \\
&\quad + c_j I_2(2)\} \qquad \text{for case 1} \qquad\qquad (1.245a)
\end{aligned}
$$

$$
\begin{aligned}
g_j &= \lim_{\rho \to 0} \frac{l_e}{2} \int_{\rho}^{2} [a_j + b_j(1 - \eta) + c_j(1 - \eta)^2] \frac{1}{4j} H_0^{(2)}\left(\frac{kl_e}{2}\eta\right) d\eta \\
&= (l_e/2)\{(a_j + b_j + c_j)I_0(2) \\
&\quad - (b_j + 2c_j)[I_1(2) - 2/(\pi k^2 l_e^2)] \\
&\quad + c_j I_2(2)\} \qquad \text{for case 2} \qquad\qquad (1.245b)
\end{aligned}
$$

$$
\begin{aligned}
g_j &= \lim_{\rho \to 0} \int_{s_1}^{s_3 - \rho} \frac{N_j}{4j} H_0^{(2)}(k|s - s_3|)\, ds \\
&\quad + \lim_{\rho \to 0} \int_{s_3 + \rho}^{s_2} \frac{N_j}{4j} H_0^{(2)}(k|s - s_3|)\, ds \\
&= \lim_{\rho \to 0} \frac{l_e}{2} \int_{-1}^{-\rho} \frac{N_j}{4j} H_0^{(2)}(k|\xi|)\, d\xi \\
&\quad + \lim_{\rho \to 0} \frac{l_e}{2} \int_{\rho}^{1} \frac{N_j}{4j} H_0^{(2)}(k|\xi|)\, d\xi \\
&= \lim_{\rho \to 0} \frac{l_e}{2} \int_{\rho}^{1} (a_j - b_j\eta + c_j\eta^2) \frac{1}{4j} H_0^{(2)}\left(\frac{kl_e}{2}\eta\right) d\eta \\
&\quad + \lim_{\rho \to 0} \frac{l_e}{2} \int_{\rho}^{1} (a_j + b_j\eta + c_j\eta^2) \frac{1}{4j} H_0^{(2)}\left(\frac{kl_e}{2}\eta\right) d\eta \\
&= l_e[a_j I_0(1) + c_j I_2(1)] \qquad \text{for case 3} \qquad\qquad (1.245c)
\end{aligned}
$$

with

$$
\begin{aligned}
I_2(\eta) &= \int \frac{\eta^2}{4j} H_0^{(2)}\left(\frac{kl_e}{2}\eta\right) d\eta \\
&= \left(\frac{2}{kl_e}\right)^2 \left[\frac{kl_e}{2}\frac{\eta^2}{4j} H_1^{(2)}\left(\frac{kl_e}{2}\eta\right) + \frac{\eta}{4j} H_0^{(2)}\left(\frac{kl_e}{2}\eta\right) - I_0(\eta)\right].
\end{aligned}
$$
$$(1.246)$$

1.7   Element Assembly

The matrix representation of each element can be carried out provided only that the node locations of that one element are known, without any knowledge of the nature of the entire mesh. Conversely, assembly of all individual element matrices to form the global matrix representation only requires knowledge of the mesh topology, i.e., of the manner in which the elements are interconnected.

To show the way of element assembly, consider the very simple mesh representation of Fig. 1.14 in which linear triangular elements are used. The three nodes of each element are identified in terms of their global node numbers and are listed in Table 1.11, where the local numbers of these nodes are defined in some consistent manner; by numbering counter-clockwise from one corner (see Fig. 1.3(a)).

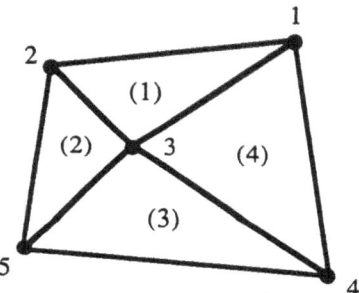

Figure 1.14: Element division.

Table 1.11: Correspondence between local and global systems.

Element	Global node numbers		
(1)	1	2	3
(2)	2	5	3
(3)	5	4	3
(4)	4	1	3
Local node numbers	1	2	3

The matrix equation for each element takes the form

$$
\begin{bmatrix}
A_{11}^{(e)} & A_{12}^{(e)} & A_{13}^{(e)} \\
A_{12}^{(e)} & A_{22}^{(e)} & A_{23}^{(e)} \\
A_{13}^{(e)} & A_{23}^{(e)} & A_{33}^{(e)}
\end{bmatrix}
\begin{bmatrix}
\phi_1^{(e)} \\
\phi_2^{(e)} \\
\phi_3^{(e)}
\end{bmatrix}
=
\begin{bmatrix}
\psi_1^{(e)} \\
\psi_2^{(e)} \\
\psi_3^{(e)}
\end{bmatrix} .
\tag{1.247}
$$

Adding the contributions of all different elements, we obtain the following global matrix equation:

$$
\left[
\begin{array}{ccc}
A_{11}^{(1)} + A_{22}^{(4)} & A_{12}^{(1)} & A_{13}^{(1)} + A_{23}^{(4)} \\
& A_{22}^{(1)} + A_{11}^{(2)} & A_{23}^{(1)} + A_{13}^{(2)} \\
& & A_{33}^{(1)} + A_{33}^{(2)} + A_{33}^{(3)} + A_{33}^{(4)} \\
& \text{sym.} &
\end{array}
\right.
$$

$$
\left.
\begin{array}{cc}
A_{12}^{(4)} & 0 \\
0 & A_{12}^{(2)} \\
A_{23}^{(3)} + A_{13}^{(4)} & A_{23}^{(2)} + A_{13}^{(3)} \\
A_{22}^{(3)} + A_{11}^{(4)} & A_{12}^{(3)} \\
& A_{22}^{(2)} + A_{11}^{(3)}
\end{array}
\right]
\begin{bmatrix}
\phi_1 \\
\phi_2 \\
\phi_3 \\
\phi_4 \\
\phi_5
\end{bmatrix}
=
\begin{bmatrix}
\psi_1^{(1)} + \psi_2^{(4)} \\
\psi_2^{(1)} + \psi_1^{(2)} \\
0 \\
\psi_2^{(3)} + \psi_1^{(4)} \\
\psi_2^{(2)} + \psi_1^{(3)}
\end{bmatrix} .
\tag{1.248}
$$

As a general rule, coefficient matrices derived from integral equations, as in the BEM, are full or nearly full. On the other hand, discretization by means of finite elements tends to produce sparse matrices and banded matrices, because any one nodal variable will be directly connected only to nodal variables which appear in the same finite element.

1.8 Imposition of Boundary Conditions

In the boundary-value problem to be solved, certain boundary segments have specified field values; thus a certain subset of the fields must assume exactly those prescribed values. Suppose the node numbering in the connected model is such that all nodes whose fields are free to vary are numbered first, all nodes with prescribed field values last. Equation (1.52) may then be written with the matrices in partitioned form

$$
\begin{bmatrix}
[A]_{11} & [A]_{12} \\
[A]_{21} & [A]_{22}
\end{bmatrix}
\begin{bmatrix}
\{\phi\}_1 \\
\{\phi\}_2
\end{bmatrix}
=
\begin{bmatrix}
\{\psi\}_1 \\
\{\psi\}_2
\end{bmatrix}
\tag{1.249}
$$

where the components of the $\{\phi\}_1$ vector in the region Ω except the boundary Γ_f, on which the Dirichlet condition is imposed, are unknown and the components of the $\{\phi\}_2$ vector on the boundary Γ_f are fixed to

$\phi = \hat{\phi}$. The components of the $\{\psi\}_1$ vector corresponding to nodes on the boundary Γ_n, on which the Neumann condition is imposed, are fixed to $\psi = \hat{\psi}$ and the other components become zero.

Using the fixed values $\{\phi\}_2 = \{\hat{\phi}\}_2$ and $\{\psi\}_1 = \{\hat{\psi}\}_1$, Eq. (1.249) becomes

$$[A]_{11}\{\phi\}_1 + [A]_{12}\{\hat{\phi}\}_2 = \{\hat{\psi}\}_1 \tag{1.250a}$$

$$[A]_{21}\{\phi\}_1 + [A]_{22}\{\hat{\phi}\}_2 = \{\psi\}_2 . \tag{1.250b}$$

From Eq. (1.250a) the unknown values of ϕ are given by

$$\{\phi\}_1 = [A]_{11}^{-1}(\{\hat{\psi}\}_1 - [A]_{12}\{\hat{\phi}\}_2) . \tag{1.251}$$

From Eqs. (1.250b) and (1.251) the unknown values of ψ are given by

$$\{\psi\}_2 = ([A]_{22} - [A]_{21}[A]_{11}^{-1}[A]_{12})\{\hat{\phi}\}_2 + [A]_{21}[A]_{11}^{-1}\{\hat{\psi}\}_1 . \tag{1.252}$$

Under the homogeneous Dirichlet boundary condition

$$\{\hat{\phi}\}_2 = \{0\} \tag{1.253}$$

Eqs. (1.251) and (1.252) become

$$\{\phi\}_1 = [A]_{11}^{-1}\{\hat{\psi}\}_1 \tag{1.254}$$

$$\{\psi\}_2 = [A]_{21}[A]_{11}^{-1}\{\hat{\psi}\}_1 . \tag{1.255}$$

When the homogeneous Neumann boundary condition

$$\{\hat{\psi}\}_1 = \{0\} \tag{1.256}$$

is assumed in addition to Eq. (1.253), from Eq. (1.249) we obtain the following eigenvalue problem:

$$[A]_{11}\{\phi\}_1 = [K]_{11}\{\phi\}_1 - k^2[M]_{11}\{\phi\}_1 = \{0\} . \tag{1.257}$$

The unknown value of ψ is given by

$$\{\psi\}_2 = [A]_{21}\{\phi\}_1 \tag{1.258}$$

where $\{\phi\}_1$ is the eigenvector for the system under consideration.

References

1) O. C. Zienkiewicz, *The Finite Element Method* (3rd edn.), McGraw-Hill, London, 1977.

2) M. V. K. Chari and P. P. Silvester (eds.), *Finite Elements in Electrical and Magnetic Field Problems*, Wiley, Chichester, 1980.

3) P. P. Silvester and R. L. Ferrari, *Finite Elements for Electrical Engineers*, Cambridge University Press, Cambridge, 1983.

4) T. Itoh (ed.), *Numerical Techniques for Microwave and Millimeter-Wave Passive Structures*, Wiley, New York, 1989.

5) E. Yamashita (ed.), *Analysis Methods for Electromagnetic Wave Problems*, Artech House, Boston, 1990.

6) M. Koshiba and M. Suzuki, "Application of the boundary-element method to waveguide discontinuities", *IEEE Trans. Microwave Theory Tech.*, Vol. MTT-34, No. 2, pp. 301–307, Feb. 1986.

Chapter 2

PLANAR OPTICAL WAVEGUIDES

2.1 Introduction

Planar optical waveguides (2-D optical waveguides) formed by diffusing transition metal into $LiNbO_3$ or $LiTaO_3$ are of considerable interest for constructing electrooptic modulators, switches, and so on. The rigorous solution is limited for special refractive index profiles, and then several methods, such as the WKB method, the perturbation method, the geometrical-optics method, the differential-numbered-solution (DNS) method, the F-matrix method, the multilayer-approximation method (MAM), the variational method, and the FEM,[1)−10)] have been developed for the other profiles. The MAM and the FEM are valid for the solution of planar anisotropic waveguides with arbitrary permittivity tensor. In the MAM, the graded variation of refractive index is approximated by homogeneous layers with a uniform refractive index in each layer and it seems difficult to consider the actual variation in each layer. In the FEM, on the other hand, the cross section of the planar waveguide is divided into the line elements and it is easy to consider the variation of refractive index in each element. Therefore, it seems that the FEM enables one to compute easily and accurately the mode spectrum of planar anisotropic waveguides with arbitrary refractive index profile.

In this chapter the FEM is described for the analysis of planar anisotropic diffused optical waveguides with a diagonal permittivity tensor. The accuracy of the method has been checked by calculating the propagation characteristics of diffused planar waveguides with exponential or Gaussian index profiles. Semiconductor-clad planar waveguides and phase-locked diode laser arrays are also analyzed.

2.2   Method of Analysis

2.2.1   Basic equations

We consider a planar anisotropic diffused optical waveguide in Fig. 2.1 having a diagonal permittivity tensor, where the region $y_1 \leq y \leq y_2$ is occupied by the inhomogeneous medium, the regions $y \leq y_1$ and $y \geq y_2$ consist of the semi-infinite substrate and the semi-infinite cover with uniform refractive index, respectively, z is taken in the direction of propagation, and there is no variation in the x direction.

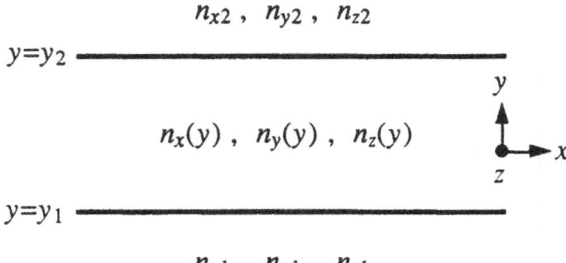

Figure 2.1: Planar optical waveguide.

With a time dependence of the form $\exp(j\omega t)$ and a z-dependence of the form $\exp(-j\beta z)$ being implied, from Maxwell's equations the following Helmholtz equation is derived:

$$\frac{d}{dy}\left(p_y \frac{d\phi}{dy}\right) - p_z \beta^2 \phi + q k_0^2 \phi = 0 \qquad (2.1)$$

with

$$k_0 = \omega\sqrt{\varepsilon_0 \mu_0} = 2\pi/\lambda \qquad (2.2)$$

where k_0 and λ are the wavenumber and wavelength of free space, respectively, and β is the phase constant in the z direction. The field ϕ and the coefficients p_y, p_z, q are written as follows:

(1) TE modes:

$$\phi = E_x \qquad (2.3)$$

$$p_y = p_z = 1 \qquad (2.4a)$$

$$q = n_x^2 \qquad (2.4b)$$

where n_x is the refractive index in the x direction and the other fields are given by

$$H_y = \frac{\beta}{k_0 Z_0} E_x \tag{2.5a}$$

$$H_z = -j \frac{1}{k_0 Z_0} \frac{dE_x}{dy} \tag{2.5b}$$

$$E_y = E_z = H_x = 0 \tag{2.5c}$$

with

$$Z_0 = \sqrt{\mu_0/\varepsilon_0} = 376.73 \; \Omega \,. \tag{2.6}$$

Here Z_0 is the impedance of free space.

(2) TM modes:

$$\phi = H_x \tag{2.7}$$

$$p_y = 1/n_z^2 \tag{2.8a}$$

$$p_z = 1/n_y^2 \tag{2.8b}$$

$$q = 1 \tag{2.8c}$$

where n_y and n_z are the refractive indices in the y and z directions, respectively, and the other fields are given by

$$E_y = -\frac{\beta Z_0}{k_0 n_y^2} H_x \tag{2.9a}$$

$$E_z = j \frac{Z_0}{k_0 n_z^2} \frac{dH_x}{dy} \tag{2.9b}$$

$$E_x = H_y = H_z = 0 \,. \tag{2.9c}$$

The functional for Eq. (2.1) is given by

$$F = \int_{y_1}^{y_2} \left(p_y \frac{d\phi^*}{dy} \frac{d\phi}{dy} + p_z \beta^2 \phi^* \phi - q k_0^2 \phi^* \phi \right) dy$$
$$+ p_{y1} \phi_1^* \psi_1 - p_{y2} \phi_2^* \psi_2 \tag{2.10}$$

where the subscripts 1 and 2 designate the values at $y = y_1$ and $y = y_2$, respectively, and $\psi = d\phi/dy$.

2.2.2 Finite element approach

Dividing the region $y_1 \leq y \leq y_2$ into a number of quadratic line elements (see Fig. 1.2(b)), we expand the field ϕ in each element as

$$\phi = \{N\}^{\mathrm{T}}\{\phi\}_e \,. \tag{2.11}$$

Substituting Eq. (2.11) into Eq. (2.10), from the variational principle (for a small admissible variation of ϕ^*, namely, $\delta\phi^*$) we obtain the following global matrix equation:

$$[K]\{\phi\} - k_0^2[M]\{\phi\} = \{\psi\} \tag{2.12}$$

with

$$[K] = \sum_e \int_e [p_y\{N_y\}\{N_y\}^{\mathrm{T}} + p_z\beta^2\{N\}\{N\}^{\mathrm{T}}]\,dy \tag{2.13}$$

$$[M] = \sum_e \int_e q\{N\}\{N\}^{\mathrm{T}}\,dy \tag{2.14}$$

$$\{\psi\} = \begin{bmatrix} -p_{y1}\psi_1 \\ 0 \\ \vdots \\ p_{y2}\psi_2 \end{bmatrix} \tag{2.15}$$

where the first and last components of the $\{\phi\}$ vector are the values of ϕ at nodes on $y = y_1$ and $y = y_2$, respectively.

2.2.3 Analytical approach

The field ϕ in the uniform region $y \leq y_1$ or $y \geq y_2$ is given by

$$\phi \propto \begin{cases} \exp(\alpha_{y1}y) & \text{for } y \leq y_1 \\ \exp(-\alpha_{y2}y) & \text{for } y \geq y_2 \end{cases} \tag{2.16}$$

with

$$\alpha_y = \sqrt{(p_z/p_y)\beta^2 - (q/p_y)k_0^2} \tag{2.17}$$

where α_y is the attenuation constant in the y direction and the subscripts 1 and 2 designate the solutions in the regions $y \leq y_1$ and $y \geq y_2$, respectively. From Eq. (2.16) the values of ψ at $y = y_1$ and $y = y_2$ are given by

$$\psi_1 = \alpha_{y1}\phi_1 \tag{2.18}$$

$$\psi_2 = -\alpha_{y2}\phi_2. \tag{2.19}$$

2.2.4 Combination of finite element and analytical relations

Substituting Eqs. (2.18) and (2.19) into Eq. (2.12), we obtain the following final matrix equation:

$$[A]\{\phi\} = \{0\} \tag{2.20}$$

with

$$[A] = [\tilde{K}] - k_0^2[M] \tag{2.21}$$

$$[\tilde{K}] = [K] + \begin{bmatrix} p_{y1}\alpha_{y1} & 0 & \cdots & 0 \\ 0 & 0 & \cdots & 0 \\ \vdots & \vdots & \ddots & \vdots \\ 0 & 0 & \cdots & p_{y2}\alpha_{y2} \end{bmatrix} . \tag{2.22}$$

The condition that Eq. (2.20) has a nontrivial solution is given by

$$|A| = 0 \tag{2.23}$$

which is the proper equation for the guided modes in a planar diffused optical waveguide.

For a waveguide composed of lossy or active media the phase constant β becomes complex and is replaced by

$$\gamma = \beta - j\alpha \tag{2.24}$$

where γ and α are the propagation and attenuation constants in the z direction, respectively. If the attenuation constant is sufficiently small, $\alpha \simeq 0$, then $\gamma \simeq \beta$, and thus β is also called the propagation constant.

Finite element approaches for the analysis of planar anisotropic diffused optical waveguides with arbitrary permittivity tensor have already been developed.[1),3)]

2.3 Planar Diffused Optical Waveguides with Film Overlays

Since the refractive index of the planar waveguide formed by diffusion changes continuously along the y direction, when evaluating Eqs. (2.13) and (2.14), numerical integration must be made or the constant index within each element must be assumed. For the sake of efficiency, the coefficients p_y, p_z, q in Eqs. (2.13) and (2.14) are defined in terms of their values at nodes:

$$p_y(y) = \{N\}^{\mathrm{T}}\{p_y\}_e \tag{2.25a}$$

$$p_z(y) = \{N\}^{\mathrm{T}}\{p_z\}_e \tag{2.25b}$$

$$q(y) = \{N\}^{\mathrm{T}}\{q\}_e \tag{2.25c}$$

where the shape function vector $\{N\}$ is the same as that used in Eq. (2.11).

Substituting Eq. (2.25) into Eqs. (2.13) and (2.14), we can integrate the expression in closed form and can avoid numerical integration. This approximation is better in accuracy and in CPU time than the approximation in which the refractive index of each element is replaced by its mean value.

First, we consider the diffused planar waveguide with dielectric or metal film as shown in Fig. 2.2, where n_c, n_f, and n_s are the refractive indices of the cover, film, and substrate, respectively, Δn is the maximum index change induced by diffusion, t is the film thickness, and the refractive index profile function is given by

$$f(y) = \exp(y/d) . \tag{2.26}$$

Here d is the diffusion depth.

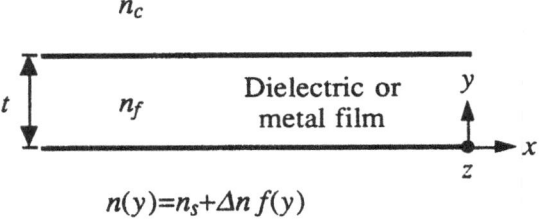

$$n(y)=n_s+\Delta n\, f(y)$$

Figure 2.2: Planar optical waveguide with film overlays.

Figures 2.3(a) and (b) show, respectively, the phase constant normalized by the free-space wavenumber, $n_{\text{eff}} = \beta/k_0$, which is called the effective index or the mode index, and the attenuation constant α (dB/cm) for the fundamental TE (solid lines) and TM (broken lines) modes as a function of film thickness, where

$$\lambda = 0.633 \ \mu\text{m} , \quad d = 0.94 \ \mu\text{m}$$
$$n_c = 1.0 , \quad n_s = 2.29 , \quad \Delta n = 0.01$$
$$n_f^2 = \begin{cases} 5.0625 & \text{for } \text{Nb}_2\text{O}_5 \\ -10.28 - j1.040 & \text{for gold} \\ -16.32 - j0.5414 & \text{for silver .} \end{cases}$$

The region divided into the line elements is $-5d \leq y \leq t$. The number of line elements in the region $-5d \leq y \leq 0$ is 25 and that in the region $0 \leq y \leq t$ is 4 to 6. The finite element solutions agree well with the exact solutions.[11),12)]

Next, we consider the diffused planar waveguide with a Gaussian index profile

$$f(y) = \exp\{-(y/d)^2\} . \tag{2.27}$$

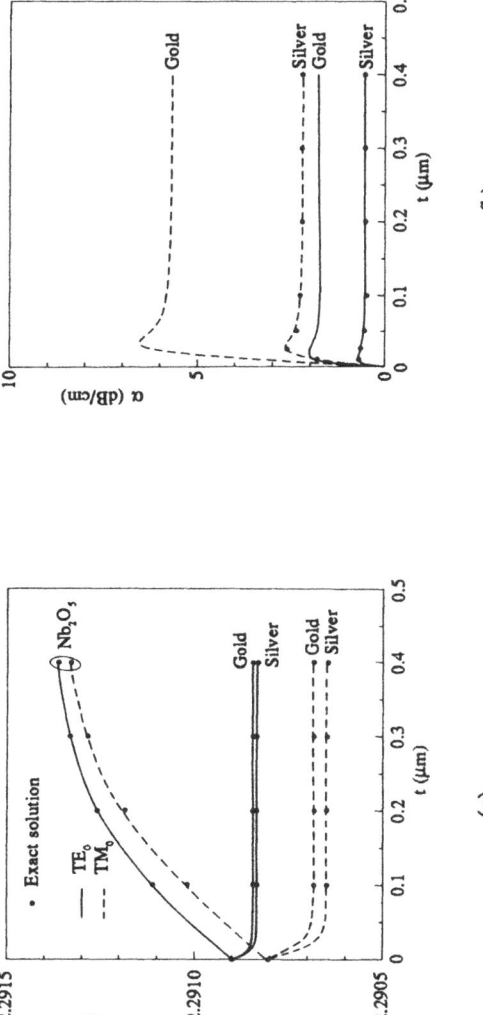

Figure 2.3: Propagation characteristics for a diffused planar waveguide with dielectric or metal film. (a) Effective index. (b) Attenuation.

In this case exact solutions are not available.

Table 2.1 shows the convergence behavior in the calculation of the effective index, where N_E is the number of elements, N_P is the number of nodes, and

$$\lambda = 0.6328 \ \mu m \,, \quad d = 5 \ \mu m \,, \quad t = 0$$
$$n_c = 1.0 \,, \quad n_s = 2.2 \,, \quad \Delta n = 0.01 \,.$$

The region divided into the line elements is $-2d \leq y \leq 0$. In this table the results of the WKB method, the multilayer-approximation method (MAM), and the differential-numbered-solution (DNS) method are also presented.[13] The finite element solutions agree well with the results of the DNS method.[13]

2.4   Semiconductor-Clad Planar Optical Waveguides

Semiconductor-clad dielectric planar waveguides exhibit many interesting properties, owing to coupling between the lossless modes of the dielectric guiding layer and the lossy modes supported by the semiconductor cladding.[14] It has also been suggested that semiconductor-clad waveguides may be useful as amplitude modulators, phase modulators, selective frequency filters, or as selective mode filters for integrated optics.[14]

The semiconductor-clad planar waveguide structure under consideration is shown in Fig. 2.4, where n_f is the index of a dielectric layer of thickness t and n_a is the index of a cladding layer of thickness h.

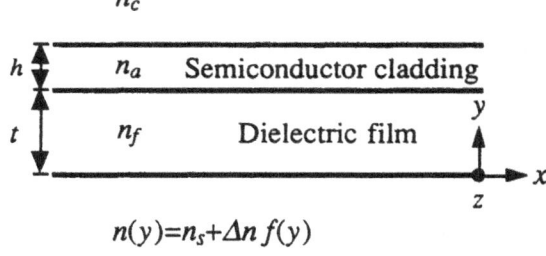

$$n(y) = n_s + \Delta n \ f(y)$$

Figure 2.4: Semiconductor-clad planar optical waveguide.

Setting artificial boundaries in the position far from the guiding region and imposing the condition

$$\phi = 0 \tag{2.28}$$

or

$$d\phi/dy = 0 \tag{2.29}$$

on the artificial boundaries, $y = y_1$ and $y = y_2$, we obtain for Eq. (2.20)

$$[K]\{\phi\} - \gamma^2[M]\{\phi\} = 0 \tag{2.30}$$

Table 2.1: Convergence behavior in calculation of the effective index for the TE modes of the planar waveguide with a Gaussian index profile.

d (μm)	Mode	WKB	MAM	DNS	FEM		
					$N_E = 10$ $N_P = 21$	$N_E = 20$ $N_P = 41$	$N_E = 30$ $N_P = 61$
3.0	TE_0	2.20577	2.20583	2.20581	2.20581	2.20581	2.20581
	TE_1	2.20155	2.20160	2.20159	2.20158	2.20158	2.20158
5.0	TE_0	2.20735	2.20738	2.20734	2.20734	2.20734	2.20734
	TE_1	2.20426	2.20430	2.20426	2.20425	2.20426	2.20426
	TE_2	2.20184	2.20187	2.20184	2.20182	2.20184	2.20184
	TE_3	2.20027	2.20027	2.20028	2.20025	2.20027	2.20027

with

$$[K] = \sum_e \int_e [q k_0^2 \{N\}\{N\}^{\mathrm{T}} - p_y \{N_y\}\{N_y\}^{\mathrm{T}}] \, dy \qquad (2.31)$$

$$[M] = \sum_e \int_e p_z \{N\}\{N\}^{\mathrm{T}} \, dy. \qquad (2.32)$$

Equation (2.30) is a standard complex eigenvalue problem whose eigenvalue and eigenvector correspond to γ^2 and $\{\phi\}$, respectively. Thus, one can solve it efficiently using a well-established eigenvalue solver. The real and imaginary parts of γ give, respectively, the phase constant and the attenuation constant, namely, $\beta = \mathrm{Re}(\gamma)$ and $\alpha = -\mathrm{Im}(\gamma)$.

First, we consider the four-layer air ($n_c = 1$)-semiconductor-dielectric ($n_f = 1.588$, $t = 1.0$ μm)-glass substrate ($n(y) = n_s = 1.51$) case reported by Batchman and McWright.[14]

Figures 2.5(a) and (b) show propagation characteristics for the TE modes of the silicon- and germanium-clad waveguides, respectively, where a free-space wavelength of 0.6328 μm is assumed, and $n_a^2 = 16.76 - j1.75$ and $n_a^2 = 14.43 - j19.54$ for silicon and germanium, respectively. The solid and dashed lines represent the effective index and the attenuation computed by the FEM, respectively. Comparison of the finite element solutions with the results of the analytical approach[14] indicated by circles and dots shows good agreement. Figure 2.5(a) shows that the attenuation and effective index of the silicon-clad waveguide behave as exponentially damped sinusoids when the silicon thickness is increased. Figure 2.5(b) shows that the larger conductivity of germanium nearly eliminates the damped oscillatory behavior in the thickness region of interest.

As shown in Fig. 2.5(a), the four-layer slab waveguide exhibits high attenuation. Next, we consider the silicon-clad diffused planar waveguide without a dielectric layer, namely, $t = 0$, where the refractive index profile of the substrate is assumed to have an exponential function profile (see Eq. (2.26)).

Figure 2.6(a) shows the computed results obtained by using the FEM, where the solid and broken lines indicate the effective index and the attenuation, respectively, the maximum index change Δn is 0.078, the diffusion depth d is 1 μm, and the cladding thickness h is varied from 0.01 to 0.5 μm. The curves behave as a damped periodic oscillation with increasing silicon thickness. The curves in Fig. 2.6(a) are almost identical to those for the four-layer air-silicon-dielectric case in Fig. 2.5(a). Since the attenuation for the diffused case in Fig. 2.6(a) is as high as that for the four-layer case,[14] the attenuation must be reduced significantly for practical devices.

The effects of the maximum index change Δn on the attenuation and effective index have also been investigated. The results for $\Delta n = 0.01$ are

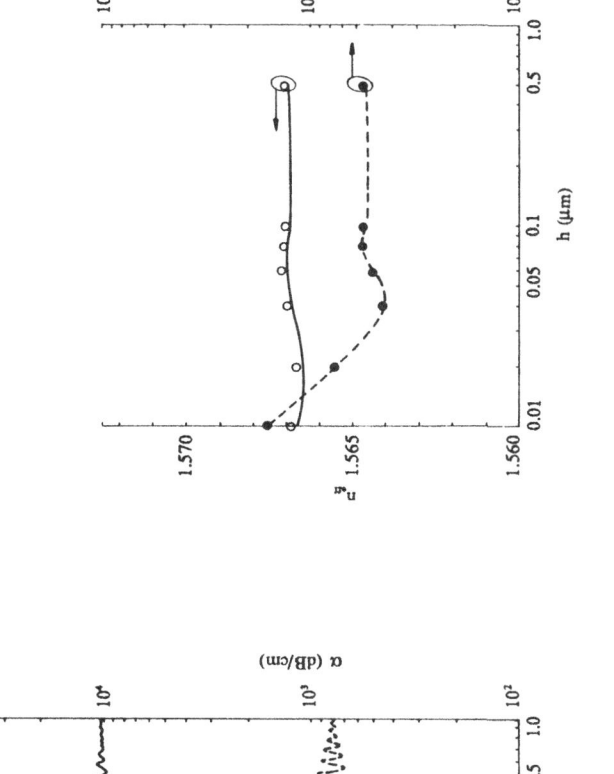

Figure 2.5: Propagation characteristics for the TE mode of a semiconductor-clad planar waveguide. (a) Silicon cladding. (b) Germanium cladding.

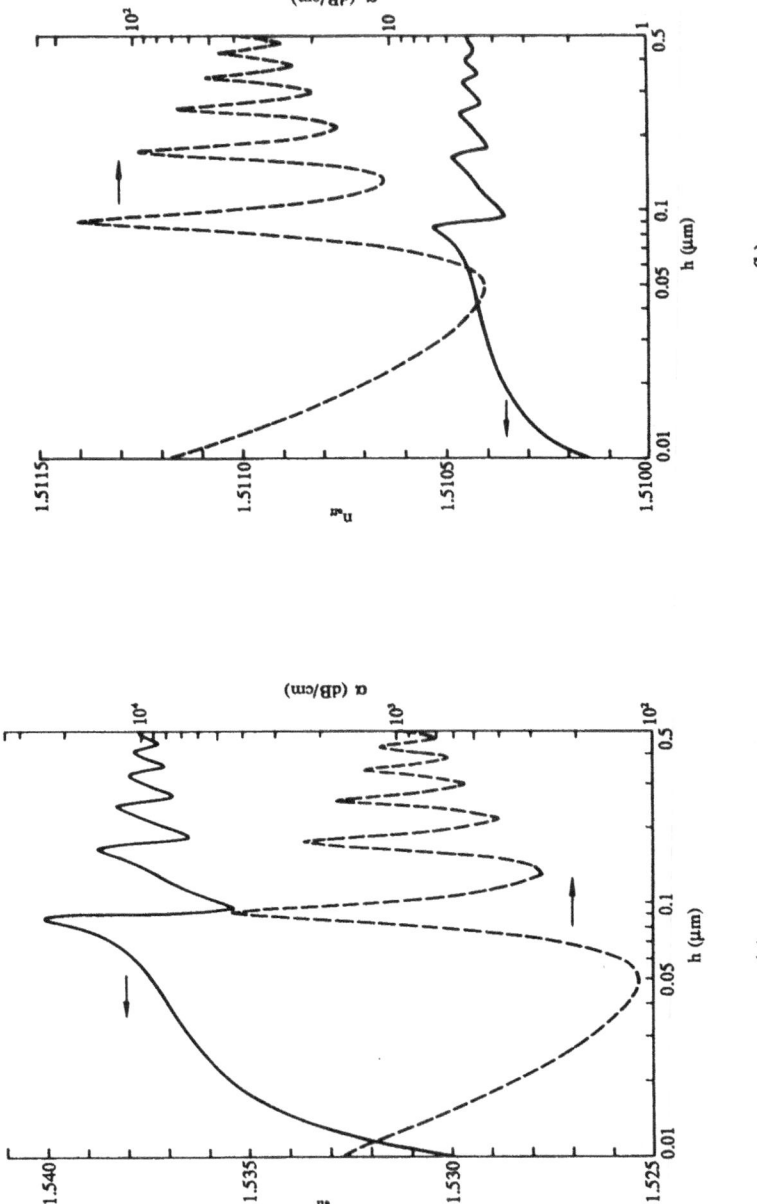

Figure 2.6: Propagation characteristics for the TE mode of a silicon-clad diffused planar waveguide. (a) $\Delta n = 0.078$.
(b) $\Delta n = 0.01$.

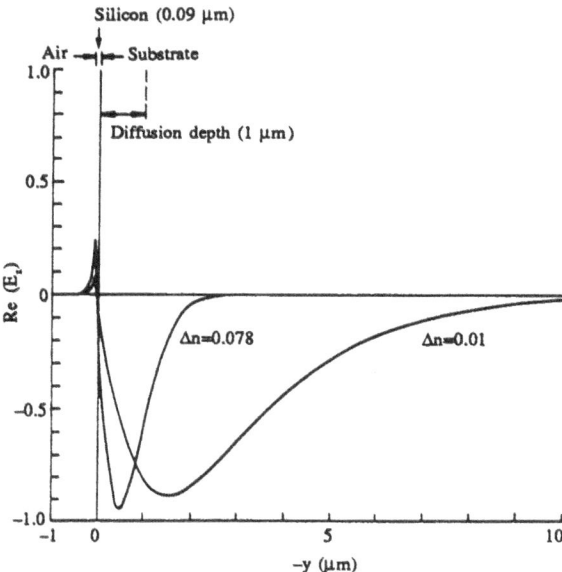

Figure 2.7: Field distributions for the TE mode of a silicon-clad diffused planar waveguide.

shown in Fig. 2.6(b). Note that the damped periodic oscillation is still present and that the attenuation is reduced significantly. It is also found that when Δn is reduced, the effective index decreases and the amplitude of the oscillation in the effective index curve also decreases.

In Fig. 2.7 the real part of the electric field E_x profile in the transverse direction is shown for a cladding thickness $h = 0.09$ μm, corresponding to the local maximum on the attenuation curves in Figs. 2.6(a) and (b). When Δn is reduced, the concentration of the field in the neighborhood of the interface between a cladding and a substrate is also reduced. We could therefore expect less attenuation when Δn is reduced.

2.5 Phase-Locked Diode Laser Arrays

Phase-locked diode laser arrays have been considered as an attractive approach to high-power semiconductor lasers having single-lobed, diffraction-limited beams.[15]

The geometry considered here is shown in Fig. 2.8; the problem is reduced to a multilayered structure ($\partial/\partial y \equiv 0$) via the effective index approximation valid for $W \gg t$, where W and t are the width and thickness of emitter, S is the spacing between adjacent emitters, n_f is the complex refractive index of the guiding region, and n_s and n_a are the complex refractive indices of the evanescent region. In the effective index method, first, assuming $W > t$, we obtain the effective index n_{TE} or n_{TM} from the characteristic equation for the TE or TM modes propagating in the slab waveguide with thickness t, respectively. Next, considering the equivalent multilayered waveguide as shown in Fig. 2.8, the following Helmholtz equation is derived:

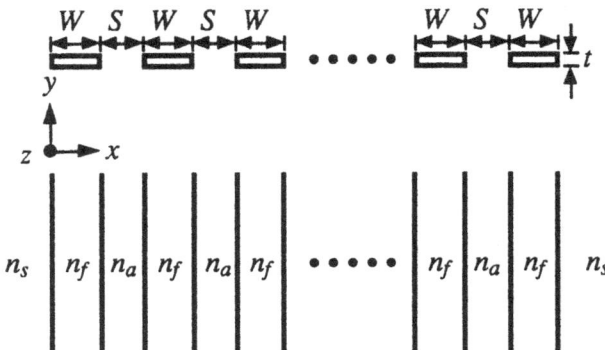

Figure 2.8: Schematic illustration of a multi-element laser array and its effective index approximation model.

$$\frac{d}{dx}\left(p\frac{d\phi}{dx}\right) - p\beta^2\phi + qk_0^2\phi = 0 \qquad (2.33)$$

where the field ϕ and the coefficients p, q are given by

$$\phi = H_y \qquad (2.34)$$
$$p = 1/n_{TE}^2 \qquad (2.35a)$$
$$q = 1 \qquad (2.35b)$$

for the TE-like modes, and

$$\phi = E_y \qquad (2.36)$$

$$p = 1 \qquad (2.37a)$$
$$q = n_{TM}^2 \qquad (2.37b)$$

for the TM-like modes. Note that Eq. (2.33) for the TE-like or TM-like modes is the same as that for the TM or TE modes propagating in the layered waveguide uniform along the y direction, respectively.

Direct application of the one-dimensional finite element technique to Eq. (2.33) yields the following matrix equation of the system:

$$[K]\{\phi\} - \gamma^2[M]\{\phi\} = \{0\} \tag{2.38}$$

with

$$[K] = \sum_e \int_e [qk_0^2\{N\}\{N\}^{\mathrm{T}} - p\{N_x\}\{N_x\}^{\mathrm{T}}]\,dx \tag{2.39}$$

$$[M] = \sum_e \int_e p\{N\}\{N\}^{\mathrm{T}}\,dx\,. \tag{2.40}$$

Equation (2.38) is a standard complex eigenvalue problem whose eigenvalue and eigenvector correspond to γ^2 and $\{\phi\}$, respectively. The real and imaginary parts of γ give, respectively, the effective index $n_{\mathrm{eff}} = \mathrm{Re}(\gamma)/k_0$ and the modal gain $g = \mathrm{Im}(\gamma)/k_0$. Furthermore, $\{\phi\}$ and its Fourier transform give, respectively, the near and far fields along the x direction of an array. The far-field intensity pattern is expressed approximately as[15]

$$I(\theta) \propto \cos^2\theta \left| \int_{-\infty}^{\infty} \phi(x)\exp(jk_0x\sin\theta)\,dx \right|^2 \tag{2.41}$$

where $\phi(x)$ is the near-field pattern (NFP) which is calculated from the FEM as the modal field, θ is the angle of observation measured in the junction plane, and $\cos\theta$ is the obliquity factor.

As a practical laser array configuration, we consider an eight-element diode laser array investigated by Kapon et al.,[15] where the refractive-index structure is illustrated in Fig. 2.9. This array supports a set of eight guided lateral modes, or supermodes. Each supermode with mode number m is characterized by its effective index n_{eff}, relative modal gain $g_r = g/[\mathrm{Im}(n_f)]$, and lateral field distribution $\phi(x)$. From the symmetrical nature of the system, only one half of the cross section is divided into 47 quadratic line elements and the following conditions are imposed on a plane of symmetry:

$$\partial\phi/\partial n = 0 \qquad \text{for } m = 1, 3, 5, \cdots \tag{2.42a}$$

$$\phi = 0 \qquad \text{for } m = 2, 4, 6, \cdots. \tag{2.42b}$$

The effective index and relative modal gain of the TE-like supermodes are listed in Table 2.2, where

$$\lambda = 1.3\ \mu\mathrm{m}\,, \quad W = 3\ \mu\mathrm{m}\,, \quad S = 2\ \mu\mathrm{m}$$
$$n_f = 3.31040 + j10^{-4}$$
$$n_s = 3.30698 - j10^{-4}$$
$$n_a = 3.30698\,.$$

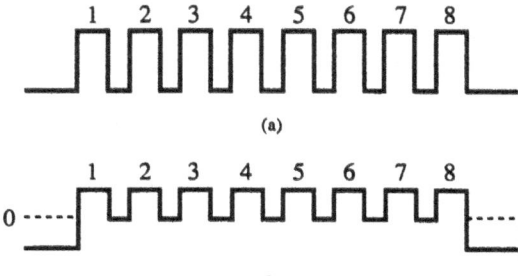

Figure 2.9: Refractive-index profile of an eight-element laser array. (a) Real part. (b) Imaginary part.

It is found from Table 2.2 that the highest-order supermode ($m = 8$) has

Table 2.2: Effective index, relative modal gain, and beam divergence of the supermodes supported by an eight-element diode laser array.

m	n_{eff}	g_r	$\Delta\theta$
1	3.30919	0.713	1.8°
2	3.30910	0.706	4.2°
3	3.30894	0.698	6.0°
4	3.30873	0.690	7.8°
5	3.30848	0.691	9.4°
6	3.30819	0.712	11.2°
7	3.30788	0.769	13.0°
8	3.30763	0.862	15.8°

the highest modal gain for the index profile given in Fig. 2.9. The lateral near-field patterns of the eight supermodes supported by this array struc-ture are shown in Fig. 2.10. It is seen that for the fundamental supermode ($m = 1$), all the coupled optical fields of the otherwise isolated channels are in phase. For the highest-order mode, on the other hand, adjacent coupled fields are 180° out of phase. The far-field patterns (FFP) of the eight supermodes are given in Fig. 2.11. These patterns show that only the fundamental supermode can produce a single-lobed diffraction limited FFP. All the other supermodes are characterized by double-lobed far fields with lobe separation that increases with increasing supermode number m. Each supermode contributes a different effective beam divergence $\Delta\theta$ to the FFP. These effective beam widths are indicated in Fig. 2.11 by the arrows and their values are also listed in Table 2.2.

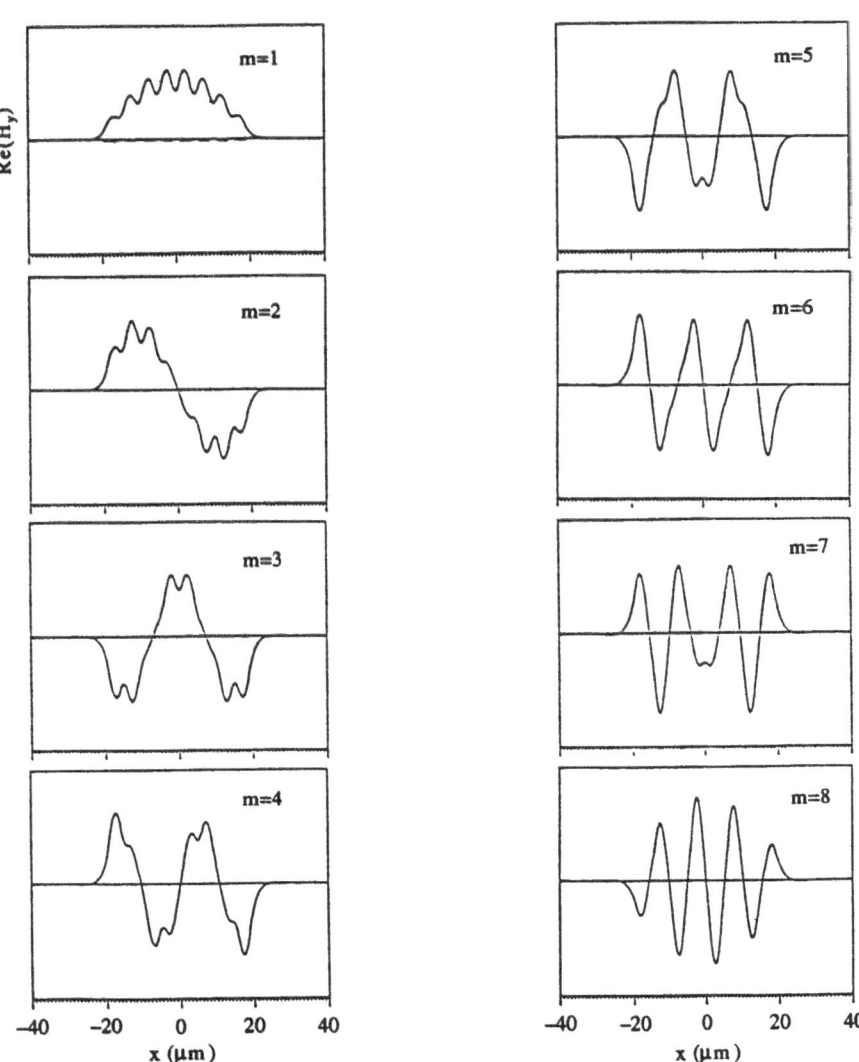

Figure 2.10: NFPs for eight supermodes.

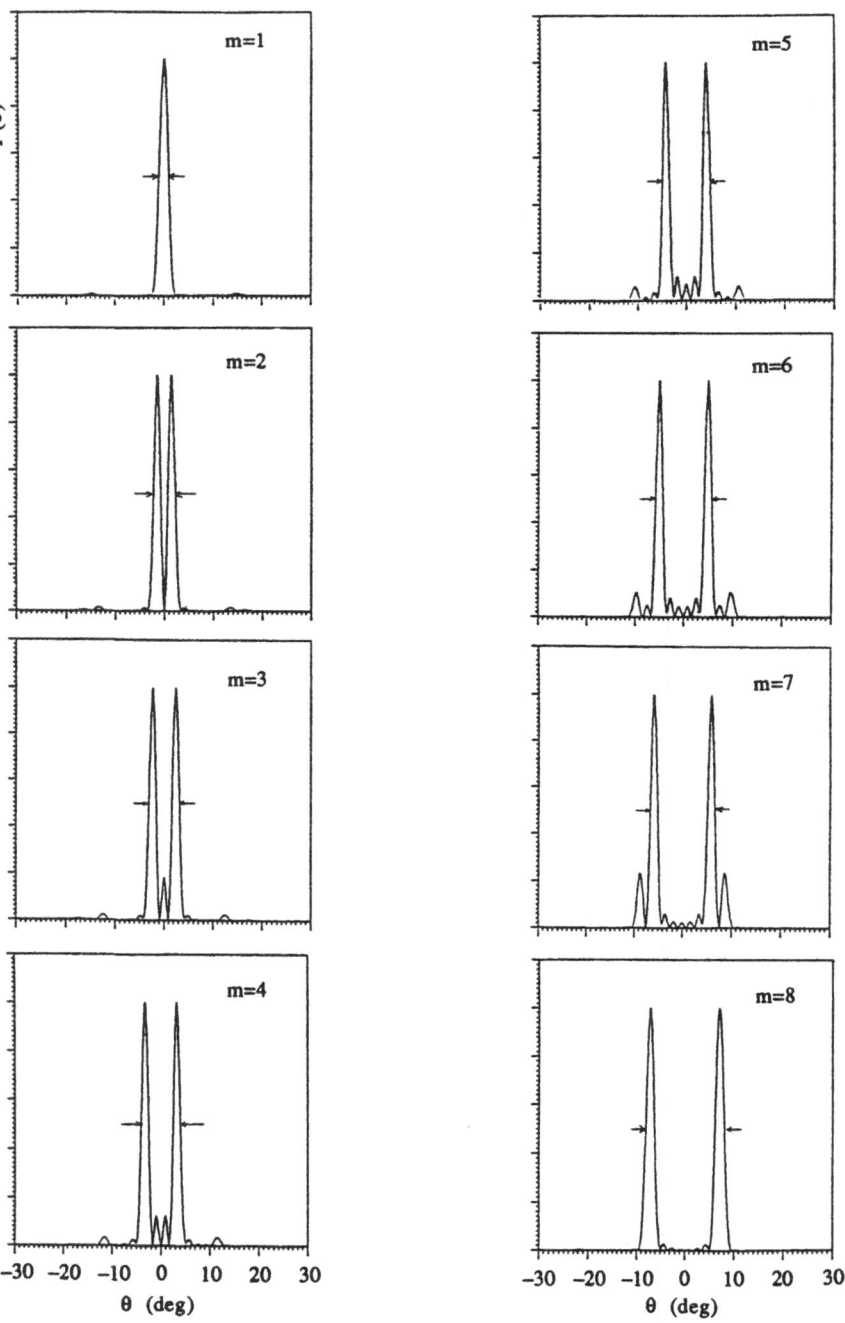

Figure 2.11: FFPs for eight supermodes.

References

1) M. Koshiba and M. Suzuki, "Numerical analysis of planar arbitrarily anisotropic diffused optical waveguides using finite-element method", *Electron. Lett.*, Vol. 18, No. 13, pp. 579–581, June 1982.

2) M. Koshiba, H. Kumagami, and M. Suzuki, "Numerical solution of silicon-clad diffused planar optical waveguides", *Electron. Lett.*, Vol. 21, No. 8, pp. 340–341, April 1985.

3) M. Koshiba, H. Kumagami, and M. Suzuki, "Finite-element solution of planar arbitrarily anisotropic diffused optical waveguides", *IEEE/OSA Jour. Lightwave Technol.*, Vol. LT-3, No. 4, pp. 773–778, Aug. 1985.

4) D. Yevick and B. Hermansson, "New fast Fourier transform and finite-element approaches to the calculation of multiple-stripe-geometry laser modes", *Jour. Appl Phys.*, Vol. 59, No. 5, pp. 1769–1771, March 1986.

5) K. Hayata and M. Koshiba, "Direct supermode analysis of phase-locked diode laser arrays", *Electron. Lett.*, Vol. 23, No. 18, pp. 935–936, Aug. 1987.

6) M. Koshiba and H. Kumagami, "Theoretical study of silicon-clad planar diffused optical waveguides", *IEE Proc.*, Vol. 134, Pt. J, No. 6, pp. 333–338, Dec. 1987.

7) B. L. Weiss and A. P. Zhao, "The influence of well shape on the propagation characteristics of multiquantum well waveguides", *IEEE Photonics Technol. Lett.*, Vol. 2, No. 11, pp. 264–268, Nov. 1990.

8) H. Kumagami and M. Koshiba, "Finite-element method using Hermitian line elements for planar diffused optical waveguides", *IEE Proc.*, Vol. 138, Pt. J, No. 3, pp. 218–220, June 1991.

9) A. Z. Zhao and S. R. Cvetkovic, "Finite element analysis of semiconductor laser arrays", *Microwave Opt. Technol. Lett.*, Vol. 4, No. 7, pp. 247–250, June 1991.

10) H. Kumagami and M. Koshiba, "Frequency response of silicon-clad planar diffused optical waveguides", *IEE Proc.*, Vol. 138, Pt. J, No. 4, pp. 249–252, Aug. 1991.

11) E. M. Conwell, "Modes in optical waveguides formed by diffusion", *Appl. Phys. Lett.*, Vol. 23, No. 6, pp. 328–329, Sept. 1973.

12) T. Findakly and C. L. Chen, "Diffused optical waveguides with exponential profile: Effects of metal-clad and dielectric overlay", *Appl. Opt.*, Vol. 17, No. 3, pp. 469–474, Feb. 1978.

13) K. Yamanouchi, T. Kamiya, and K. Shibayama, "New leaky surface waves in anisotropic metal-diffused optical waveguides", *IEEE Trans. Microwave Theory Tech.*, Vol. MTT-26, No. 4, pp. 298–305, April 1978.

14) T. E. Batchman and G. M. McWright, "Mode coupling between dielectric and semiconductor planar waveguides", *IEEE Jour. Quantum Electron.*, Vol. QE-18, No. 4, pp. 782–788, April 1982.

15) E. Kapon, Z. Rav-noy, S. Margalit, and A. Yariv, "Phase-locked arrays of buried-ridge InP/InGaAsP diode lasers", *IEEE/OSA Jour. Lightwave Technol.*, Vol. LT-4, No. 7, pp. 919–925, July 1986.

Chapter 3

OPTICAL CHANNEL WAVEGUIDES

3.1 Introduction

To rigorously evaluate propagation characteristics of optical channel waveguides (3-D optical waveguides) with arbitrarily shaped cross section, vectorial wave analysis is necessary, and different types of vector FEMs, according to the components of electromagnetic fields, are used:

1. FEM using axial components of E and H fields,[1)−22)]
2. FEM using full vector E or H field,[23)−62)]
3. FEM using both E and H fields,[63)−66)]
4. FEM using transverse components of E or H field,[67)−73)]
5. FEM using transverse components of both E and H fields,[74)−79)] and
6. FEM using vector and scalar potentials.[80),81)]

Although various two-dimensional, finite element formulations have been developed to date, spurious, nonphysical solutions appear in the vectorial wave analysis.

Method 1 is standard for vectorial wave analysis and has been utilized for various waveguides in microwave and millimeter wave regions. Recently, it has also been utilized for the waveguide for use in optical wavelength region. In this approach, however, spurious solutions appear in the slow-wave region ($n_{\text{eff}} \geq 1$), and what is worse, these are coupled with physical solutions. This results in a great difficulty for the analysis of optical waveguides whose guided modes exist in the slow-wave region. We must discriminate the physical solutions from the spurious solutions by their field profiles. Consequently, the development of a method to suppress or eliminate such spurious solutions is pressingly needed, and research on this topic has been extensive in recent years. Of the various formulations, method 2 is quite suitable for a wide range of practical, complicated problems. This approach has been widely used for the solution of various waveguiding structures in microwave, millimeter-wave, and optical wavelength regions, and recently has been utilized as the waveguide solver of CAD packages.[39),43),52)]

In this chapter the FEM in terms of three components of the electric or magnetic field is described for the vectorial wave analysis of optical channel waveguides, and the approximate scalar FEM (SFEM)[82)−95)] is also introduced. The SFEM has as its main advantages: smaller matrix dimensions, less computer time, and no spurious solutions.

3.2   Method of Vectorial Wave Analysis

3.2.1   Basic equations

We consider an optical channel waveguide in Fig. 3.1 having a diagonal permittivity tensor $[\varepsilon_r]$.

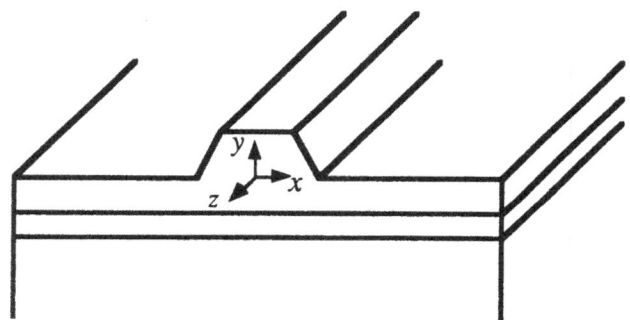

Figure 3.1: Optical channel waveguide.

With a time dependence of the form $\exp(j\omega t)$ and a z-dependence of the form $\exp(-j\beta z)$ being implied, from Maxwell's equations the following vectorial wave equation is derived:

$$\nabla \times ([\varepsilon_r]^{-1}\nabla \times \boldsymbol{H}) - k_0^2 \boldsymbol{H} = 0 \,. \qquad (3.1)$$

The functional for Eq. (3.1) is given by

$$F = \iint_{\Omega} [(\nabla \times \boldsymbol{H})^* \cdot ([\varepsilon_r]^{-1}\nabla \times \boldsymbol{H}) - k_0^2 \boldsymbol{H}^* \cdot \boldsymbol{H}] \, dx \, dy \qquad (3.2)$$

where $\iint_{\Omega} dx \, dy$ denotes the integration over the waveguide cross section Ω. Because the magnetic field is continuous anywhere in dielectric waveguides, it is convenient to use the above variational expression in terms of full vector $\boldsymbol{H}$ field for the finite element analysis.

For the functional, Eq. (3.2), the first variation δF is given by

$$\delta F = \iint_\Omega \delta \boldsymbol{H}^* \cdot [\nabla \times ([\varepsilon_r]^{-1} \nabla \times \boldsymbol{H}) - k_0^2 \boldsymbol{H}] \, dx \, dy$$

$$- \int_\Gamma \delta \boldsymbol{H}^* \cdot [\boldsymbol{n} \times ([\varepsilon_r]^{-1} \nabla \times \boldsymbol{H})] \, d\Gamma \qquad (3.3)$$

where Γ represents the contour of the region Ω, $\boldsymbol{n}$ is the outward unit normal vector to Γ, and the term $\boldsymbol{n} \times ([\varepsilon_r]^{-1} \nabla \times \boldsymbol{H})$ corresponds to the tangential components of the electric field $\boldsymbol{E}$ on Γ. The stationary requirement shows that

$$\nabla \times ([\varepsilon_r]^{-1} \nabla \times \boldsymbol{H}) - k_0^2 \boldsymbol{H} = 0 \quad \text{in region } \Omega \qquad (3.4)$$

as the Euler equation which is coincident with the wave equation, Eq. (3.1), and

$$\boldsymbol{n} \times ([\varepsilon_r]^{-1} \nabla \times \boldsymbol{H}) = 0 \quad \text{on perfect electric conductor (PEC)} \qquad (3.5)$$

as the natural boundary condition, since $\delta \boldsymbol{H}^*$ in Eq. (3.3) is arbitrary. The forced boundary condition is given by

$$\boldsymbol{n} \times \boldsymbol{H} = 0 \quad \text{on perfect magnetic conductor (PMC).} \qquad (3.6)$$

3.2.2 Finite element approach

Dividing the waveguide cross section into a number of quadratic triangular elements (see Fig. 1.3(b)), we expand the magnetic field $\boldsymbol{H}$ in each element as

$$\boldsymbol{H} = [N]^{\mathrm{T}} \{H\}_e \qquad (3.7)$$

with

$$\{H\}_e = \begin{bmatrix} \{H_x\}_e \\ \{H_y\}_e \\ \{H_z\}_e \end{bmatrix} \qquad (3.8)$$

$$[N] = \begin{bmatrix} \{N\} & \{0\} & \{0\} \\ \{0\} & \{N\} & \{0\} \\ \{0\} & \{0\} & j\{N\} \end{bmatrix} \qquad (3.9)$$

where $\{H_x\}_e$, $\{H_y\}_e$ and $\{H_z\}_e$ are the nodal magnetic-field vectors.

Substituting Eq. (3.7) into Eq. (3.2), from the variational principle we obtain the following eigenvalue problem:

$$[K]\{H\} - k_0^2 [M]\{H\} = \{0\} \qquad (3.10)$$

with

$$[K] = \begin{bmatrix} [K_{xx}] & [K_{xy}] & [K_{xz}] \\ [K_{yx}] & [K_{yy}] & [K_{yz}] \\ [K_{zx}] & [K_{zy}] & [K_{zz}] \end{bmatrix}$$

$$= \sum_e \iint_e [B]^* [\varepsilon_r]^{-1} [B]^T dx\, dy \tag{3.11}$$

$$[M] = \begin{bmatrix} [M_{xx}] & [0] & [0] \\ [0] & [M_{yy}] & [0] \\ [0] & [0] & [M_{zz}] \end{bmatrix}$$

$$= \sum_e \iint_e [N]^* [N]^T dx\, dy \tag{3.12}$$

$$[B] = \begin{bmatrix} \{0\} & -j\beta\{N\} & -\{N_y\} \\ j\beta\{N\} & \{0\} & \{N_x\} \\ j\{N_y\} & -j\{N_x\} & \{0\} \end{bmatrix} \tag{3.13}$$

where $[0]$ is a null matrix, and the submatrices of $[K]$ and $[M]$ are given by

$$[K_{xx}] = \sum_e \iint_e \left[\frac{\beta^2}{n_y^2}\{N\}\{N\}^T + \frac{1}{n_z^2}\{N_y\}\{N_y\}^T \right] dx\, dy \tag{3.14a}$$

$$[K_{xy}] = [K_{yx}]^T = \sum_e \iint_e \left[-\frac{1}{n_z^2}\{N_y\}\{N_x\}^T \right] dx\, dy \tag{3.14b}$$

$$[K_{xz}] = [K_{zx}]^T = \sum_e \iint_e \frac{\beta}{n_y^2}\{N\}\{N_x\}^T dx\, dy \tag{3.14c}$$

$$[K_{yy}] = \sum_e \iint_e \left[\frac{\beta^2}{n_x^2}\{N\}\{N\}^T + \frac{1}{n_z^2}\{N_x\}\{N_x\}^T \right] dx\, dy \tag{3.14d}$$

$$[K_{yz}] = [K_{zy}]^T = \sum_e \iint_e \frac{\beta}{n_x^2}\{N\}\{N_y\}^T dx\, dy \tag{3.14e}$$

$$[K_{zz}] = \sum_e \iint_e \left[\frac{1}{n_x^2}\{N_y\}\{N_y\}^T + \frac{1}{n_y^2}\{N_x\}\{N_x\}^T \right] dx\, dy$$

$$\tag{3.14f}$$

$$[M_{xx}] = [M_{yy}] = [M_{zz}] = \sum_e \iint_e \{N\}\{N\}^T dx\, dy . \tag{3.15}$$

3.3 Spurious Solutions

3.3.1 Cause of spurious solutions

Spurious solutions also appear in the finite element analysis using the full vector $\boldsymbol{H}$ field. The spurious solutions encountered in Eq. (3.10) do not satisfy the condition

$$\nabla \cdot \boldsymbol{H} = 0 \,. \tag{3.16}$$

These spurious solutions fall into two fairly clear categories. The first one (S_1) can be characterized as follows:

$$\nabla \times \boldsymbol{H} = 0 \,, \quad \nabla \cdot \boldsymbol{H} \neq 0 \quad \text{for} \quad k_0^2 = 0 \,. \tag{3.17}$$

The second group (S_2) can be characterized as follows:

$$\nabla \times \boldsymbol{H} \neq 0 \,, \quad \nabla \cdot \boldsymbol{H} \neq 0 \quad \text{for}_{,} \quad k_0^2 > 0 \,. \tag{3.18}$$

Now, we discuss the cause of spurious solutions.

It is obvious that for zero eigenvalues ($k_0^2 = 0$), Eq. (3.4) does not guarantee the satisfaction of the condition $\nabla \cdot \boldsymbol{H} = 0$.

In the FEM, piecewise defined polynomial functions are used for trial functions, and the unknown function and its derivatives are continuous and discontinuous at the element interface, respectively. (Note that finite element solutions are obtained to satisfy the continuity conditions for tangential electric-field components at the interface between elements 1 and 2, namely, $\boldsymbol{n}_1 \times ([\varepsilon_{r1}]^{-1} \nabla \times \boldsymbol{H}_1) = -\boldsymbol{n}_2 \times ([\varepsilon_{r2}]^{-1} \nabla \times \boldsymbol{H}_2)$.) Therefore, for the magnetic field $\boldsymbol{H}$ calculated from the FEM, the trivial relation

$$\nabla \cdot \{\nabla \times ([\varepsilon_r]^{-1} \nabla \times \boldsymbol{H})\} = 0 \tag{3.19}$$

is not always satisfied. This means that, for nonzero eigenvalues ($k_0^2 > 0$), the satisfaction of the condition $\nabla \cdot \boldsymbol{H} = 0$ is not guaranteed either. In fact, the derivatives of $\boldsymbol{H}$ do abruptly change at element interfaces, and thus the components of $\boldsymbol{H}$ rapidly oscillate over the whole region. After all, to satisfy the vector formula, Eq. (3.19), it is necessary to consider the condition $\nabla \cdot \boldsymbol{H} = 0$ as a constraint.

Actually, it was confirmed in the 1980s that the spurious solutions encountered in Eq. (3.10) do not satisfy the condition $\nabla \cdot \boldsymbol{H} = 0$. Soon afterward, it was expected that the spurious solutions could be eliminated if the condition $\nabla \cdot \boldsymbol{H} = 0$ was considered, and several approaches, in which the condition $\nabla \cdot \boldsymbol{H} = 0$ was explicitly imposed, were proposed.

3.3.2   Penalty function method

The penalty function method[25)−28),30)−35),38),39),43)−45),47),50)−53),55)] has been studied extensively and applied to various waveguiding problems in recent years because it can suppress or eliminate the spurious solutions. In this approach, the variational expression, Eq. (3.2), is modified into

$$F = \iint_\Omega [(\nabla \times \boldsymbol{H})^* \cdot ([\varepsilon_r]^{-1} \nabla \times \boldsymbol{H}) - k_0^2 \boldsymbol{H}^* \cdot \boldsymbol{H}] \, dx \, dy$$

$$+ s \iint_\Omega (\nabla \cdot \boldsymbol{H})^* (\nabla \cdot \boldsymbol{H}) \, dx \, dy \,. \tag{3.20}$$

Equation (3.20) is a variational expression satisfying the condition $\nabla \cdot \boldsymbol{H} = 0$ in a least squares way, where s is a positive constant called the penalty coefficient.

For the functional, Eq. (3.20), the first variation δF is given by

$$\delta F = \iint_\Omega \delta \boldsymbol{H}^* \cdot [\nabla \times ([\varepsilon_r]^{-1} \nabla \times \boldsymbol{H}) - s\nabla(\nabla \cdot \boldsymbol{H}) - k_0^2 \boldsymbol{H}] \, dx \, dy$$

$$- \int_\Gamma \delta \boldsymbol{H}^* \cdot [\boldsymbol{n} \times ([\varepsilon_r]^{-1} \nabla \times \boldsymbol{H}) - \boldsymbol{n}(\nabla \cdot \boldsymbol{H})] \, d\Gamma \,. \tag{3.21}$$

The stationary requirement shows that

$$\nabla \times ([\varepsilon_r]^{-1} \nabla \times \boldsymbol{H}) - s\nabla(\nabla \cdot \boldsymbol{H}) - k_0^2 \boldsymbol{H} = 0 \tag{3.22}$$

as the Euler equation and

$$\boldsymbol{n} \times ([\varepsilon_r]^{-1} \nabla \times \boldsymbol{H}) = 0 \qquad \text{on PEC} \tag{3.23a}$$

$$\boldsymbol{n}(\nabla \cdot \boldsymbol{H}) = 0 \qquad \text{on PMC} \tag{3.23b}$$

as the natural boundary conditions. The forced boundary conditions are given by

$$\boldsymbol{n} \times \boldsymbol{H} = 0 \qquad \text{on PMC} \tag{3.24a}$$

$$\boldsymbol{n} \cdot \boldsymbol{H} = 0 \qquad \text{on PEC} \,. \tag{3.24b}$$

Multiplying Eq. (3.22) by $\boldsymbol{H}^*$ and integrating over the region Ω, the following equation is obtained using Green's formula and the boundary conditions on Γ:

$$\iint_\Omega [(\nabla \times \boldsymbol{H})^* \cdot ([\varepsilon_r]^{-1} \nabla \times \boldsymbol{H}) + s(\nabla \cdot \boldsymbol{H})^* (\nabla \cdot \boldsymbol{H})$$

$$- k_0^2 \boldsymbol{H}^* \cdot \boldsymbol{H}] \, dx \, dy = 0 \,. \tag{3.25}$$

In Eq. (3.25), if $[\varepsilon_r]^{-1}$ is a positive definite matrix, then $\nabla \times \boldsymbol{H} = 0$ and $\nabla \cdot \boldsymbol{H} = 0$ are satisfied for $k_0^2 = 0$. Therefore, the spurious solutions S_1 are eliminated.

Taking divergence of Eq. (3.22), we obtain

$$(s\nabla^2 + k_0^2)(\nabla \cdot \boldsymbol{H}) = 0 . \tag{3.26}$$

If the curl of $\boldsymbol{H}$ is not zero for $k_0^2 > 0$, the eigenvalues k_0^2 of Eq. (3.22) cannot satisfy Eq. (3.26). Therefore, the eigenvectors of Eq. (3.22) should obey $\nabla \cdot \boldsymbol{H} = 0$, and the spurious solutions S_2 are eliminated.

When $\nabla \times \boldsymbol{H} = 0$ for $k_0^2 > 0$, Eq. (3.20) may have solutions other than those of Eq. (3.2). This new group (S_3) characterized by

$$\nabla \times \boldsymbol{H} = 0 , \qquad \nabla \cdot \boldsymbol{H} \neq 0 \qquad \text{for } k_0^2 > 0 \tag{3.27}$$

obeys the following equations:

$$\boldsymbol{H} = \nabla \phi \tag{3.28a}$$
$$(s\nabla^2 + k_0^2)\phi = 0 \qquad \text{in region } \Omega \tag{3.28b}$$
$$\partial\phi/\partial n = 0 \qquad \text{on PEC} \tag{3.28c}$$
$$\phi = 0 \qquad \text{on PMC} \tag{3.28d}$$

where ϕ is the scalar field. The magnetic field $\boldsymbol{H}$ of Eq. (3.28) satisfies the stationary requirement $\delta F = 0$, but the divergence of $\boldsymbol{H}$ is not zero. Therefore, in the finite element analysis using Eq. (3.20), the spurious solution S_3 which are not included in Eq. (3.2) do appear. Fortunately, unlike other spurious solutions encountered in the FEM, we need not worry about these solutions because, as found from Eq. (3.28), we can suppress them into the following region:

$$n_{\text{eff}} < 1/\sqrt{s} . \tag{3.29}$$

Equation (3.29) indicates that the larger is s, the narrower the region for the spurious solutions. For example, if we want to suppress the spurious solutions from the region $n_{\text{eff}} \geq n_{\text{min}}$, we have only to choose the value as $s \geq 1/n_{\text{min}}^2$. In this connection, for $s = 1$, the spurious solutions can be completely eliminated from the slow-wave region. Therefore, this approach is useful, particularly for the analysis of dielectric waveguides.

Application of the standard FEM to Eq. (3.20) yields

$$([K] + s[L])\{H\} - k_0^2[M]\{H\} = 0 \tag{3.30}$$

with

$$[L] = \begin{bmatrix} [L_{xx}] & [L_{xy}] & [L_{xz}] \\ [L_{yx}] & [L_{yy}] & [L_{yz}] \\ [L_{zx}] & [L_{zy}] & [L_{zz}] \end{bmatrix}$$

$$= \sum_e \iint_e \{C\}\{C\}^{\mathrm{T}}\, dx\, dy \tag{3.31}$$

$$\{C\} = \left[\begin{array}{c} \{N_x\} \\ \{N_y\} \\ \beta\{N\} \end{array} \right] \tag{3.32}$$

where submatrices of $[L]$ are given by

$$[L_{xx}] = \sum_e \iint_e \{N_x\}\{N_x\}^{\mathrm{T}}\, dx\, dy \tag{3.33a}$$

$$[L_{xy}] = [L_{yx}]^{\mathrm{T}} = \sum_e \iint_e \{N_x\}\{N_y\}^{\mathrm{T}}\, dx\, dy \tag{3.33b}$$

$$[L_{xz}] = [L_{zx}]^{\mathrm{T}} = \sum_e \iint_e \beta\{N_x\}\{N\}^{\mathrm{T}}\, dx\, dy \tag{3.33c}$$

$$[L_{yy}] = \sum_e \iint_e \{N_y\}\{N_y\}^{\mathrm{T}}\, dx\, dy \tag{3.33d}$$

$$[L_{yz}] = [L_{zy}]^{\mathrm{T}} = \sum_e \iint_e \beta\{N_y\}\{N\}^{\mathrm{T}}\, dx\, dy \tag{3.33e}$$

$$[L_{zz}] = \sum_e \iint_e \beta^2\{N\}\{N\}^{\mathrm{T}}\, dx\, dy\ . \tag{3.33f}$$

Finite element approaches using the penalty function method for the analysis of optical channel waveguides with an arbitrary permittivity tensor have already been developed.[28),30),35)]

3.3.3 Transverse magnetic-field method

In the method transforming the finite element equation in terms of the full vector H field, Eq. (3.10), into that in terms of only transverse magnetic-field components, using the condition $\nabla \cdot H = 0$, the spurious solutions can be completely eliminated.[36),40),42),48),49)]

The key point of this approach is that the condition $\nabla \cdot H = 0$; that is,

$$H_z = \frac{1}{j\beta} \left(\frac{\partial H_x}{\partial x} + \frac{\partial H_y}{\partial y} \right) \tag{3.34}$$

is discretized into the following form:

$$[D_z]\{H_z\} = [D_t]\{H_t\} \tag{3.35}$$

via the Galerkin procedure. Here, $\{H_t\}$ is the nodal transverse magnetic-field vector, and the matrices $[D_z]$ and $[D_t]$ are given by

$$[D_z] = \sum_e \iint_e \{N\}\{N\}^{\mathrm{T}} \, dx \, dy \tag{3.36}$$

$$[D_t] = \sum_e \iint_e \left[-\frac{1}{\beta}\{N\}\{N_x\}^{\mathrm{T}} - \frac{1}{\beta}\{N\}\{N_y\}^{\mathrm{T}} \right] dx \, dy . \tag{3.37}$$

Using Eq. (3.35), the global magnetic-field vector $\{H\}$ is expressed as

$$\{H\} = [D]\{H_t\} \tag{3.38}$$

with

$$[D] = \left[\begin{array}{c} [1] \\ [D_z]^{-1}[D_t] \end{array} \right] \tag{3.39}$$

where [1] is a unit matrix.

Considering Eq. (3.38) into the finite element equation in terms of the full vector H field, Eq. (3.10), the following finite element equation in terms of transverse magnetic-field components can be obtained:

$$[K_{tt}]\{H_t\} - k_0^2[M_{tt}]\{H_t\} = \{0\} \tag{3.40}$$

with

$$[K_{tt}] = [D]^{\mathrm{T}}[K][D] \tag{3.41}$$

$$[M_{tt}] = [D]^{\mathrm{T}}[M][D] . \tag{3.42}$$

In Eq. (3.40), the spurious solutions are completely eliminated, and the matrix size is two-thirds that of Eqs. (3.10) and (3.30). Moreover, an artificial coefficient such as the one included in the penalty function method (the accuracy of the solution depends on the penalty coefficient s in the penalty function method) is not included at all. Since the matrices $[K_{tt}]$ and $[M_{tt}]$ are dense, unlike those generated by the penalty function approach, there is no possibility of taking advantage of sparsity with a sophisticated eigenvalue solver.

3.3.4   Direct solution for propagation constants

Because the eigenvalues of the FEM using the full vector H field correspond to k_0^2, in practical calculations, β is first given as an input datum, and subsequently k_0 is obtained as a solution. Therefore, iterations are required if we want to deal with a medium whose material constant varies with k_0 (dispersive medium) or a medium with loss or gain (complex medium). To avoid such a difficulty, a method that can obtain β directly by giving k_0 as an input datum has recently been developed.[42),49)]

Equation (3.30) may be rewritten as

$$[K]\{H\} + \beta[L]\{H\} + \beta^2[M]\{H\} = \{0\} \tag{3.43}$$

with

$$[K] = \begin{bmatrix} [K_{xx}] & [K_{xy}] & [0] \\ [K_{yx}] & [K_{yy}] & [0] \\ [0] & [0] & [K_{zz}] \end{bmatrix} \tag{3.44}$$

$$[L] = \begin{bmatrix} [0] & [0] & [L_{xz}] \\ [0] & [0] & [L_{yz}] \\ [L_{zx}] & [L_{zy}] & [0] \end{bmatrix} \tag{3.45}$$

$$[M] = \begin{bmatrix} [M_{xx}] & [0] & [0] \\ [0] & [M_{yy}] & [0] \\ [0] & [0] & [M_{zz}] \end{bmatrix} \tag{3.46}$$

where submatrices of $[K]$, $[L]$, and $[M]$ are given by

$$[K_{xx}] = \sum_e \iint_e \left[\frac{1}{n_z^2}\{N_y\}\{N_y\}^{\mathrm{T}} + s\{N_x\}\{N_x\}^{\mathrm{T}} - k_0^2\{N\}\{N\}^{\mathrm{T}} \right] dx\,dy \tag{3.47a}$$

$$[K_{xy}] = [K_{yx}]^{\mathrm{T}} = \sum_e \iint_e \left[-\frac{1}{n_z^2}\{N_y\}\{N_x\}^{\mathrm{T}} + s\{N_x\}\{N_y\}^{\mathrm{T}} \right] dx\,dy \tag{3.47b}$$

$$[K_{yy}] = \sum_e \iint_e \left[\frac{1}{n_z^2}\{N_x\}\{N_x\}^{\mathrm{T}} + s\{N_y\}\{N_y\}^{\mathrm{T}} - k_0^2\{N\}\{N\}^{\mathrm{T}} \right] dx\,dy \tag{3.47c}$$

$$[K_{zz}] = \sum_e \iint_e \left[\frac{1}{n_x^2}\{N_y\}\{N_y\}^{\mathrm{T}} + \frac{1}{n_y^2}\{N_x\}\{N_x\}^{\mathrm{T}} - k_0^2\{N\}\{N\}^{\mathrm{T}} \right] dx\,dy \tag{3.47d}$$

$$[L_{xz}] = [L_{zx}]^{\mathrm{T}} = \sum_e \iint_e \left[\frac{1}{n_y^2}\{N\}\{N_x\}^{\mathrm{T}} + s\{N_x\}\{N\}^{\mathrm{T}} \right] dx\,dy \tag{3.48a}$$

$$[L_{yz}] = [L_{zy}]^{\mathrm{T}} = \sum_e \iint_e \left[\frac{1}{n_x^2}\{N\}\{N_y\}^{\mathrm{T}} + s\{N_y\}\{N\}^{\mathrm{T}} \right] dx\,dy \tag{3.48b}$$

$$[M_{xx}] = \sum_e \iint_e \frac{1}{n_y^2} \{N\}\{N\}^\mathrm{T}\, dx\, dy \qquad (3.49a)$$

$$[M_{yy}] = \sum_e \iint_e \frac{1}{n_x^2} \{N\}\{N\}^\mathrm{T}\, dx\, dy \qquad (3.49b)$$

$$[M_{zz}] = \sum_e \iint_e s\{N\}\{N\}^\mathrm{T}\, dx\, dy . \qquad (3.49c)$$

Since Eq. (3.43) is a quadratic eigenvalue problem, it can be reduced to the following linearized form:

$$\begin{bmatrix} [0] & [1] \\ -[M]^{-1}[K] & -[M]^{-1}[L] \end{bmatrix} \begin{bmatrix} \{H\} \\ \beta\{H\} \end{bmatrix} = \beta \begin{bmatrix} \{H\} \\ \beta\{H\} \end{bmatrix} . \qquad (3.50)$$

This form is a standard eigenvalue problem whose eigenvalue directly corresponds to the phase constant β. For the waveguides composed of lossy or active media the phase constant β is replaced by the propagation constant γ (see Eq. (2.24)). Note that Eq. (3.50) involves $6N_P$ unknown components in each eigenvector, where N_P is the number of nodes.

3.3.5 Edge element method

The use of edge elements (see Subsection 1.6.2) provides a direct solution for the propagation constant and avoids spurious solutions.[62] In the edge element method[29),41),62)] the nodal parameters are not limited to the magnetic field as in the penalty function method or in the transverse magnetic-field method.

From Maxwell's equations the following vectorial wave equation is derived:

$$\nabla \times ([p]\nabla \times \phi) - k_0^2[q]\phi = 0 \qquad (3.51)$$

with

$$[p] = \begin{bmatrix} p_x & 0 & 0 \\ 0 & p_y & 0 \\ 0 & 0 & p_z \end{bmatrix} \qquad (3.52)$$

$$[q] = \begin{bmatrix} q_x & 0 & 0 \\ 0 & q_y & 0 \\ 0 & 0 & q_z \end{bmatrix} \qquad (3.53)$$

where ϕ denotes either $\boldsymbol{E}$ or $\boldsymbol{H}$, and the components of $[p]$ and $[q]$ are given by

$$p_x = p_y = p_z = 1 ,$$

$$q_x = n_x^2, \quad q_y = n_y^2, \quad q_z = n_z^2 \quad \text{for } \phi = \boldsymbol{E} \qquad (3.54)$$

$$p_x = 1/n_x^2, \quad p_y = 1/n_y^2, \quad p_z = 1/n_z^2,$$

$$q_x = q_y = q_z = 1 \quad \text{for } \phi = \boldsymbol{H} . \qquad (3.55)$$

The functional for Eq. (3.51) is given by

$$F = \iint_\Omega [(\nabla \times \phi)^* \cdot ([p]\nabla \times \phi) - k_0^2[q]\phi^* \cdot \phi]\, dx\, dy . \qquad (3.56)$$

Dividing the waveguide cross section into a number of edge elements (see Fig. 1.8), we expand the transverse-field components ϕ_x, ϕ_y and the axial-field component ϕ_z in each element as

$$\phi = [N]^{\mathrm{T}}\{\phi\}_e \qquad (3.57)$$

with

$$\{\phi\}_e = \begin{bmatrix} \{\phi_t\}_e \\ \{\phi_z\}_e \end{bmatrix} \qquad (3.58)$$

$$[N] = \begin{bmatrix} \{U\} & \{V\} & \{0\} \\ \{0\} & \{0\} & j\{N\} \end{bmatrix} \qquad (3.59)$$

where the shape function vectors $\{U\}$ and $\{V\}$ for the edge element are given in Eqs. (1.168) and (1.169), respectively, and $\{N\}$ is the ordinary shape function vector for the linear triangular element (see Table 1.2).

Substituting Eq. (3.57) into Eq. (3.56), from the variational principle we obtain the following eigenvalue problem:

$$[K]\{\phi\} - k_0^2[M]\{\phi\} = \{0\} \qquad (3.60)$$

with

$$[K] = \begin{bmatrix} [K_{tt}] & [K_{tz}] \\ [K_{zt}] & [K_{zz}] \end{bmatrix}$$

$$= \sum_e \iint_e [B]^*[p][B]^{\mathrm{T}}\, dx\, dy \qquad (3.61)$$

$$[M] = \begin{bmatrix} [M_{tt}] & [0] \\ [0] & [M_{zz}] \end{bmatrix}$$

$$= \sum_e \iint_e [N]^*[q][N]^{\mathrm{T}}\, dx\, dy \qquad (3.62)$$

$$[B] = \begin{bmatrix} j\beta\{V\} & -j\beta\{U\} & -\{U_y\} + \{V_x\} \\ j\{N_y\} & -j\{N_x\} & \{0\} \end{bmatrix} \qquad (3.63)$$

where the submatrices of $[K]$ and $[M]$ are given by

$$[K_{tt}] = \sum_e \iint_e [p_x \beta^2 \{V\}\{V\}^T + p_y \beta^2 \{U\}\{U\}^T$$
$$+ 4p_z \{U_y\}\{U_y\}^T]\, dx\, dy \qquad (3.64a)$$

$$[K_{tz}] = [K_{zt}]^T$$
$$= \sum_e \iint_e [p_x \beta \{V\}\{N_y\}^T + p_y \beta \{U\}\{N_x\}^T]\, dx\, dy \quad (3.64b)$$

$$[K_{zz}] = \sum_e \iint_e [p_x \{N_y\}\{N_y\}^T + p_y \{N_x\}\{N_x\}^T]\, dx\, dy \quad (3.64c)$$

$$[M_{tt}] = \sum_e \iint_e [q_x \{U\}\{U\}^T + q_y \{V\}\{V\}^T]\, dx\, dy \qquad (3.65a)$$

$$[M_{zz}] = \sum_e \iint_e q_z \{N\}\{N\}^T\, dx\, dy. \qquad (3.65b)$$

Equation (3.60) may be rewritten as

$$[K_{tt}]\{\phi_t\} - \beta[K_{tz}]\{\phi_z\} - \beta^2[M_{tt}]\{\phi_t\} = \{0\} \qquad (3.66a)$$
$$-\beta[K_{zt}]\{\phi_t\} + [K_{zz}]\{\phi_z\} = \{0\} \qquad (3.66b)$$

with

$$[K_{tt}] = \sum_e \iint_e [q_x k_0^2 \{U\}\{U\}^T + q_y k_0^2 \{V\}\{V\}^T$$
$$- 4p_z \{U_y\}\{U_y\}^T]\, dx\, dy \qquad (3.67a)$$

$$[K_{tz}] = [K_{zt}]^T$$
$$= \sum_e \iint_e [p_x \{V\}\{N_y\}^T + p_y \{U\}\{N_x\}^T]\, dx\, dy \qquad (3.67b)$$

$$[K_{zz}] = \sum_e \iint_e [q_z k_0^2 \{N\}\{N\}^T - p_x \{N_y\}\{N_y\}^T$$
$$- p_y \{N_x\}\{N_x\}^T]\, dx\, dy \qquad (3.67c)$$

$$[M_{tt}] = \sum_e \iint_e [p_x \{V\}\{V\}^T + p_y \{U\}\{U\}^T]\, dx\, dy. \qquad (3.68)$$

Substituting Eq. (3.66b) into Eq. (3.66a), we obtain the following eigenvalue problem:

$$[K_{tt}]\{\phi_t\} - \beta^2[\tilde{M}_{tt}]\{\phi_t\} = \{0\} \qquad (3.69)$$

with

$$[\tilde{M}_{tt}] = [M_{tt}] + [K_{tz}][K_{zz}]^{-1}[K_{zt}] \,. \tag{3.70}$$

Note that Eq. (3.69) will give a solution directly for the propagation constant and the corresponding field distribution, and involves only the edge variables in the transversal plane.

3.4   Assessment of Vector Finite Element Method

As is well known, an optical channel waveguide will support the propagation of waves having two possible field configurations, classified as the E^x and E^y modes.[96] The main field components of the E^x_{mn} modes are E_x and H_y, while those of the E^y_{mn} modes are E_y and H_x. The subscripts m and n designate the number of maxima of the dominant field in the x and y directions, respectively.

In this section we present the propagation characteristics for the E^x and E^y modes of dielectric rectangular waveguides, equilateral triangular core waveguides, and anisotropic embedded rectangular waveguides. Numerical results are obtained by the edge element method, namely, Eq. (3.69).

First, we consider a dielectric rectangular waveguide in Fig. 3.2, where n_f and n_s are the refractive indices of the core and substrate regions, respectively. Because of the twofold symmetry of the system, we subdivide only one-quarter of the waveguide cross section into edge elements. For simplicity, assuming the artificial boundaries $x = \pm X/2$ and $y = \pm Y/2$ far from the core region, the original unbounded structure is replaced by a corresponding bounded one. Here, these artificial boundaries are assumed to be PECs. Planes of symmetry become the PEC or PMC according to the kind of modes. Conditions on the planes of symmetry, $x = 0$ and $y = 0$, are summarized in Table 3.1. Figure 3.3 shows the propagation characteristics of this waveguide, where $W = 2t$, $X = 10t$, $Y = 5t$, the number of elements $N_E = 320$, the number of corner points $N_C = 187$, the number of side points $N_S = 506$, and the normalized frequency v and the normalized propagation constant b are defined as

$$v = k_0 t \sqrt{n_f^2 - n_s^2}/\pi \tag{3.71}$$

$$b = \frac{n_{\text{eff}}^2 - n_s^2}{n_f^2 - n_s^2} = \frac{(\beta/k_0)^2 - n_s^2}{n_f^2 - n_s^2} \,. \tag{3.72}$$

The finite element solutions of edge element formulation agree well with the results of the point matching method (PMM).[97] The results of the Marcatili method (MM)[96] deviate from those of the PMM at lower frequencies.

Table 3.1: Conditions on planes of symmetry for a dielectric rectangular waveguide.

Conditions on planes of symmetry		E^x mode		E^y mode	
$x = 0$	$y = 0$	m	n	m	n
$E_y = E_z = H_x = 0$	$E_y = H_x = H_z = 0$	Odd	Odd	Even	Even
$E_x = H_y = H_z = 0$	$E_y = H_x = H_z = 0$	Even	Odd	Odd	Even
$E_y = E_z = H_x = 0$	$E_x = E_z = H_y = 0$	Odd	Even	Even	Odd
$E_x = H_y = H_z = 0$	$E_x = E_z = H_y = 0$	Even	Even	Odd	Odd

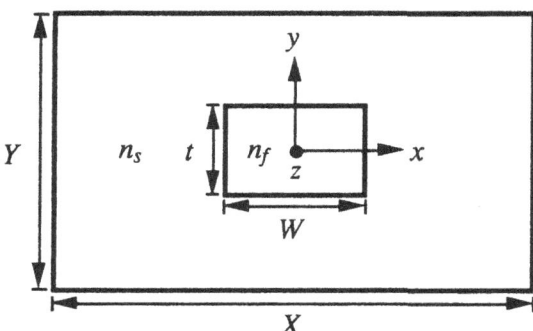

Figure 3.2: Dielectric rectangular waveguide.

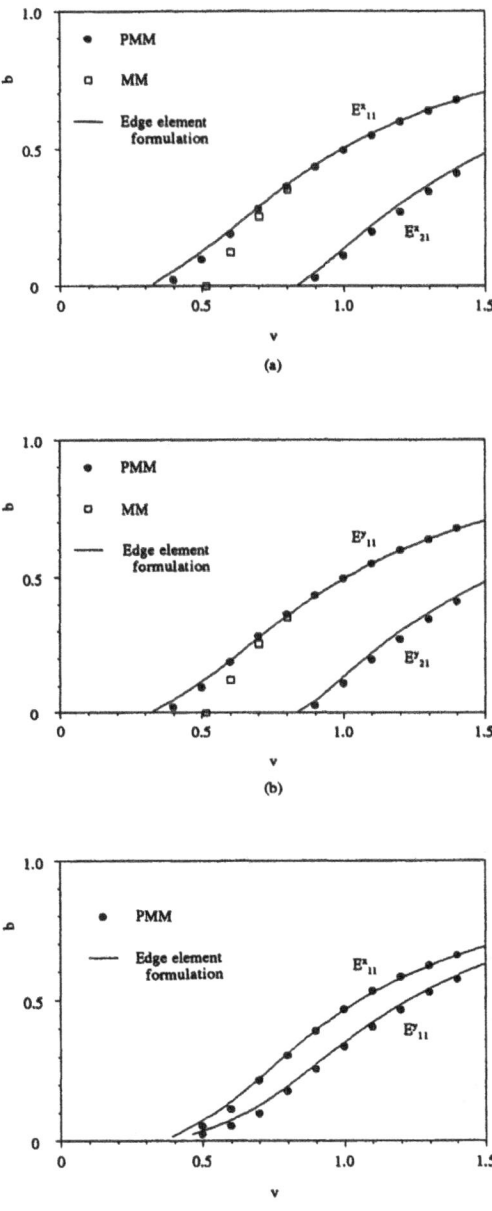

Figure 3.3: Propagation characteristics of a dielectric rectangular waveguide. (a) E_{11}^x and E_{21}^x modes (n_f=1.05, n_s=1.0). (b) E_{11}^y and E_{21}^y modes (n_f=1.05, n_s=1.0). (c) E_{11}^x and E_{11}^y modes (n_f=1.5, n_s=1.0).

Next, we consider an equilateral triangular core waveguide in Fig. 3.4 and subdivide only one-half of the waveguide cross section into edge elements. Figure 3.5 shows the propagation characteristics for the E_{11}^y mode of this waveguide, where $X = 6t$, $Y = 5t$, $N_E = 360$, $N_C = 208$, and $N_S = 567$. The finite element solutions of edge element formulation agree well with those of full vector H-field formulation with the penalty coefficient $s = 1$[34) and of axial-field (E_z and H_z) formulation.[12)

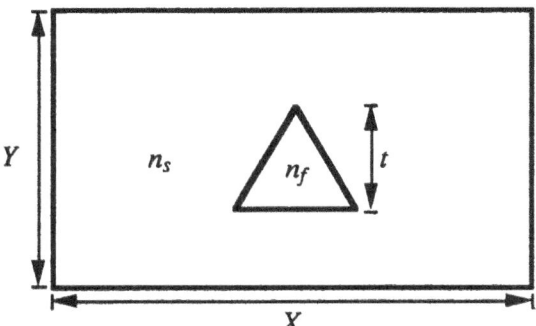

Figure 3.4: Equilateral triangular core waveguide.

Lastly, we consider an anisotropic embedded rectangular waveguide in Fig. 3.6 and subdivide only one-half of the waveguide cross section into edge elements. Figure 3.7 shows the propagation characteristics for the E_{11}^y mode of this waveguide, where $W = 5t$, $X = 10t$, $Y = 5t$, $N_E = 280$, $N_C = 165$, and $N_S = 444$. The finite element solutions of edge element formulation agree well with those of axial field formulation.[6)

Note that the spurious solutions are included in the finite element solutions of axial field formulation. To avoid confusion, such spurious solutions are not shown in Figs. 3.5(a) and 3.7. In the edge element method spurious solutions do not appear anywhere. The convergence of solutions has been checked by increasing the number of elements and the values of X and Y so that the influence of the artificial boundaries is negligible.

3.5 Method of Approximate Scalar Wave Analysis

3.5.1 Basic equations

We consider an optical channel waveguide as shown in Fig. 3.1 having a diagonal permittivity tensor.

Noting that the E^x and E^y modes are well approximated by the TE^y ($E_y \equiv 0$, a leading function is E^x) and TM^y ($H_y \equiv 0$, a leading function is H^x) modes, respectively, from Maxwell's equations the following Helmholtz

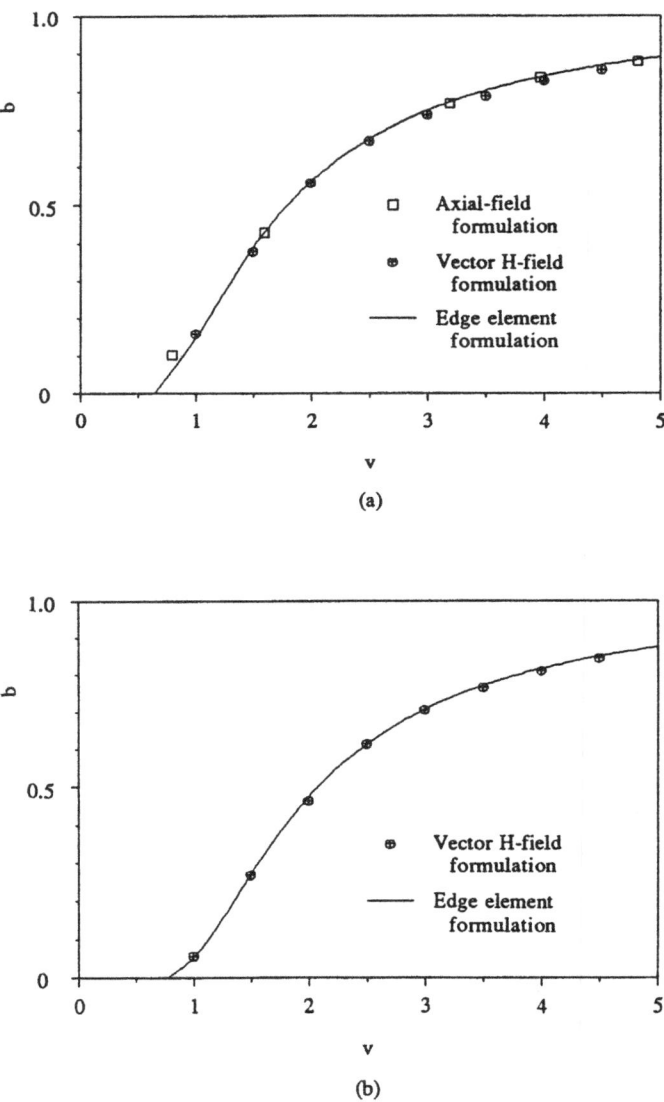

Figure 3.5: Propagation characteristics for the E_{11}^y mode of an equilateral triangular core waveguide. (a) n_f=1.5085 and n_s=1.50. (b) n_f=1.5 and n_s=1.0.

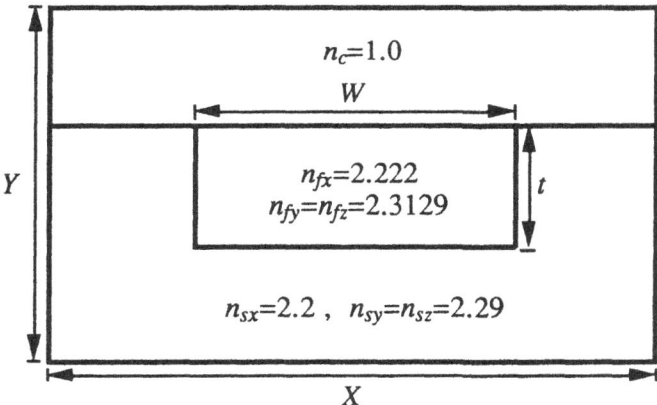

Figure 3.6: Anisotropic embedded rectangular waveguide.

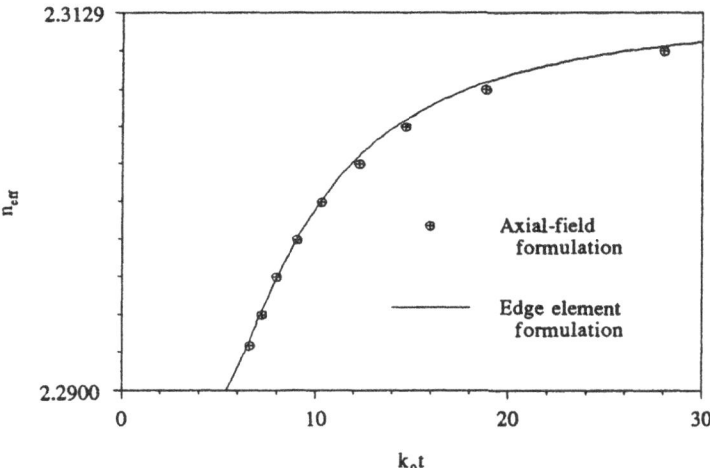

Figure 3.7: Propagation characteristics for the E_{11}^y mode of an anisotropic embedded rectangular waveguide.

equation is derived:

$$p_x \frac{\partial^2 \phi}{\partial x^2} + p_y \frac{\partial^2 \phi}{\partial y^2} - p_z \beta^2 \phi + q k_0^2 \phi = 0 \qquad (3.73)$$

where the field ϕ and the coefficients p_x, p_y, p_z, q are written as follows:

(1) E^x modes:

$$\phi = E_x \qquad (3.74)$$
$$p_x = n_x^2/n_z^2 \qquad (3.75a)$$
$$p_y = 1 \qquad (3.75b)$$
$$p_z = 1 \qquad (3.75c)$$
$$q = n_x^2 \qquad (3.75d)$$

and the remaining field components are given by

$$E_y = 0 \qquad (3.76a)$$

$$E_z = -j\frac{n_x^2}{n_z^2 \beta} \frac{\partial \phi}{\partial x} \qquad (3.76b)$$

$$H_x = \frac{n_x^2}{n_z^2 Z_0 \beta k_0} \frac{\partial^2 \phi}{\partial x \partial y} \qquad (3.76c)$$

$$H_y = \frac{1}{Z_0 \beta k_0} \left(\beta^2 \phi - \frac{n_x^2}{n_z^2} \frac{\partial^2 \phi}{\partial x^2} \right) \qquad (3.76d)$$

$$H_z = -j\frac{1}{Z_0 k_0} \frac{\partial \phi}{\partial y} \qquad (3.76e)$$

where Z_0 is the impedance of free space (see Eq. (2.6)).

(2) E^y modes:

$$\phi = H_x \qquad (3.77)$$
$$p_x = 1/n_y^2 \qquad (3.78a)$$
$$p_y = 1/n_z^2 \qquad (3.78b)$$
$$p_z = 1/n_y^2 \qquad (3.78c)$$
$$q = 1 \qquad (3.78d)$$

and the remaining field components are given by

$$E_x = -\frac{Z_0}{n_x^2 \beta k_0} \frac{\partial^2 \phi}{\partial x \partial y} \qquad (3.79a)$$

$$E_y = -\frac{Z_0}{n_y^2 \beta k_0} \left(\beta^2 \phi - \frac{\partial^2 \phi}{\partial x^2} \right) \qquad (3.79b)$$

$$E_z = j\frac{Z_0}{n_z^2 k_0}\frac{\partial \phi}{\partial y} \tag{3.79c}$$

$$H_y = 0 \tag{3.79d}$$

$$H_z = -j\frac{1}{\beta}\frac{\partial \phi}{\partial x} . \tag{3.79e}$$

The functional for Eq. (3.73) is given by

$$F = \iint_\Omega \left(p_x \frac{\partial \phi^*}{\partial x}\frac{\partial \phi}{\partial x} + p_y \frac{\partial \phi^*}{\partial y}\frac{\partial \phi}{\partial y} + p_z \beta^2 \phi^* \phi - q k_0^2 \phi^* \phi \right) dx\, dy . \tag{3.80}$$

3.5.2 Finite element approach

Dividing the waveguide cross section into a number of quadratic triangular elements (see Fig. 1.3(b)), we expand the field ϕ in each element as

$$\phi = \{N\}^{\mathrm{T}}\{\phi\}_e . \tag{3.81}$$

Substituting Eq. (3.81) into Eq. (3.80), from the variational principle we obtain the following eigenvalue problems:

$$[K]\{\phi\} - \beta^2 [M]\{\phi\} = \{0\} \tag{3.82}$$

with

$$[K] = \sum_e \iint_e [qk_0^2\{N\}\{N\}^{\mathrm{T}} - p_x\{N_x\}\{N_x\}^{\mathrm{T}}$$

$$- p_y\{N_y\}\{N_y\}^{\mathrm{T}}]\, dx\, dy \tag{3.83}$$

$$[M] = \sum_e \iint_e p_z\{N\}\{N\}^{\mathrm{T}}\, dx\, dy . \tag{3.84}$$

Approximate scalar finite element approaches for the analysis of optical channel waveguides with arbitrary permittivity tensor have already been developed.[84]

3.6 Assessment of Scalar Finite Element Method

Rib waveguides have received much attention recently due to advances in III-V semiconductor technology. The simplicity of fabrication and the ease of control of lateral and vertical index profiles make rib waveguide structures attractive for integrated optical applications.

In this section the SFEM will be applied to semiconductor rib waveguides, and it is demonstrated that the SFEM can predict accurate results and would be useful to the general user community as a practical tool of modeling and analysis for the purpose of device design and fabrication.

3.6.1 Semiconductor rib waveguides

We consider a semiconductor rib waveguide and subdivide only one-half of the waveguide cross section into quadratic triangular elements as shown in Fig. 3.8. Waveguide parameters are listed in Table 3.2.

Table 3.2: Semiconductor rib waveguide structures.

Structure	λ (μm)	n_f	n_s	n_c	W (μm)	t (μm)	h (μm)	X_s (μm)	Y_s (μm)	Y_c (μm)
1	1.55	3.44	3.34	1.0	2.0	0.2	1.1	3.0	5.0	1.0
2	1.55	3.44	3.36	1.0	3.0	0.9	0.1	3.0	5.0	1.0
3	1.55	3.44	3.435	1.0	4.0	3.5	2.5	4.5	7.5	1.0
4	1.15	3.44	3.40	1.0	3.0	$h + t = 1.0$		3.0	5.0	1.0
5	1.55	3.167	3.162	1.0	6.0	2.5	1.5	8.0	10.0	2.0
6	1.15	3.44	3.40	1.0	3.0	0.5	0.5	6.0	7.0	2.0

Table 3.3 shows the normalized propagation constant b (see Eq. (3.72)) for the E_{11}^x mode of structures 1, 2, 3, where 224 elements and 485 nodes are used for structures 1, 2, and 240 elements and 519 nodes for structure 3. In this table the results of the effective index method (EIM),[98] the equivalent network method (ENM),[99] the beam propagation method (BPM),[100] the scalar variational method (SVM),[98] and the scalar finite difference method (SFDM),[101] are also presented. Table 3.4 shows the effective index n_{eff} for the E_{11}^x and E_{11}^y modes of structure 4, where 224 elements and 485 nodes are used. In this table the results of the vector finite element method (VFEM)[33] and the SFDM[101] are also presented. The results of the SFEM agree well with those of the VFEM and of the SFDM.

One feature of the SFEM is the ease with which it will generate far-field radiation patterns. Far-field patterns are of particular interest when one wishes to couple the input or output of a laser or waveguide to another

Table 3.3: Normalized propagation constants for the E_{11}^x mode of rib waveguide structures 1, 2, 3.

Structure	EIM	ENM	BPM	SVM	SFDM	SFEM
1	0.4995	0.4782	0.5093	0.5008	0.4656	0.4981
2	0.4407	0.4390	0.4471	0.4332	0.4401	0.4380
3	0.4783	0.3728	0.2998	0.3446	0.3621	0.3974

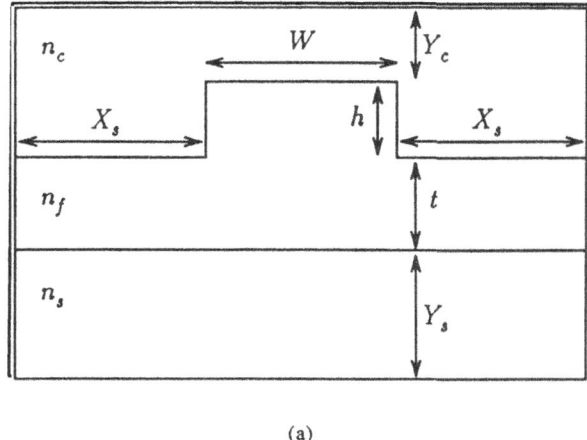

(a)

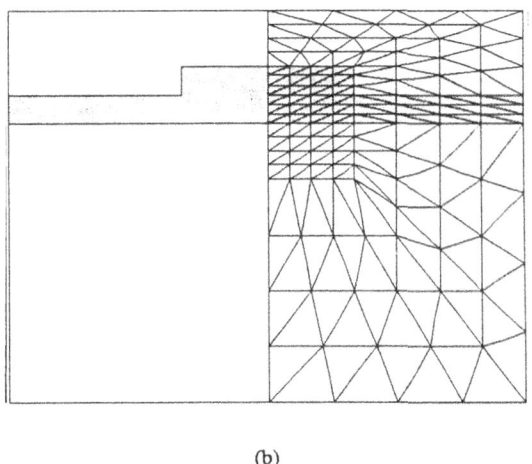

(b)

Figure 3.8: Rib waveguide. (a) Waveguide structure. (b) Element division.

Table 3.4: Effective indices for the E_{11}^x and E_{11}^y modes of rib waveguide structure 4.

t	E_{11}^x			E_{11}^y	
(μm)	VFEM	SFDM	SFEM	SFDM	SFEM
0	3.4121	3.41188	3.41204	3.41051	3.41028
0.1	3.4122	3.41200	3.41214	3.41060	3.41037
0.2	3.41235	3.41217	3.41229	3.41073	3.41051
0.3	3.41255	3.41240	3.41249	3.41092	3.41070
0.4	3.41285	3.41271	3.41276	3.41117	3.41097
0.5	3.41315	3.41310	3.41311	3.41150	3.41132
0.6	3.41365	3.41358	3.41353	3.41190	3.41174
0.7	3.4141	3.41415	3.41404	3.41241	3.41227
0.8	3.41475	3.41485	3.41468	3.41303	3.41293
0.9	3.4156	3.41568	3.41553	3.41385	3.41383

device, such as an optical fiber, and the SFEM is particularly well suited to tackling this problem. Using the near-field pattern (NFP) calculated from the SFEM as the modal field, $\phi(x, y)$, the far-field intensity pattern (FFP) is expressed approximately as[102]

$$I(\theta_x, \theta_y) \propto \cos^2 \theta \left| \iint_\Omega \phi(x, y) \exp[j(k_x x + k_y y)] dx \, dy \right|^2 \qquad (3.85)$$

with

$$k_x \quad = \quad k_0 \sin \theta_x = k_0 \sin \theta \cos \theta_a \qquad (3.86a)$$
$$k_y \quad = \quad k_0 \sin \theta_y = k_0 \sin \theta \sin \theta_a \qquad (3.86b)$$
$$\cos^2 \theta \quad = \quad 1 - (\sin^2 \theta_x + \sin^2 \theta_y) \qquad (3.86c)$$

where θ is the polar angle to the z direction, θ_a is the corresponding azimuthal angle, θ_x and θ_y are the components of θ resolved along x and y, and $\cos \theta$ is the obliquity factor.

Figures 3.9 and 3.10 show the NFP and FFP for the E_{11}^x (symmetric) mode of structure 5, respectively, where 249 elements and 538 nodes are used. Figures 3.11 and 3.12 show the NFP and FFP for the E_{21}^x (antisymmetric) mode of structure 6, respectively, where 240 elements and 521 nodes are used. The contour lines of NFP are at $\pm 10\%$, $\pm 20\%$, etc. of the peak amplitude, while those of FFP are at 10%, 20% etc. of the peak intensity. In Figs. 3.10 and 3.12 the FFPs calculated from the EIM[102] and the spectral index method (SIM)[102] are also presented. Table 3.5 shows the linewidth, namely, the full width at half maximum intensity (FWHM) evaluated from Figs. 3.10 and 3.12. In this table the peak positions for the E_{21}^x mode are also listed. The FWHM predicted by the SFEM agree approximately with that predicted by the SIM. It is found from Fig. 3.12(b) that the low intensity contours do not have the same major axis as the high intensity contours, indicating the nonseparable nature of the FFP.

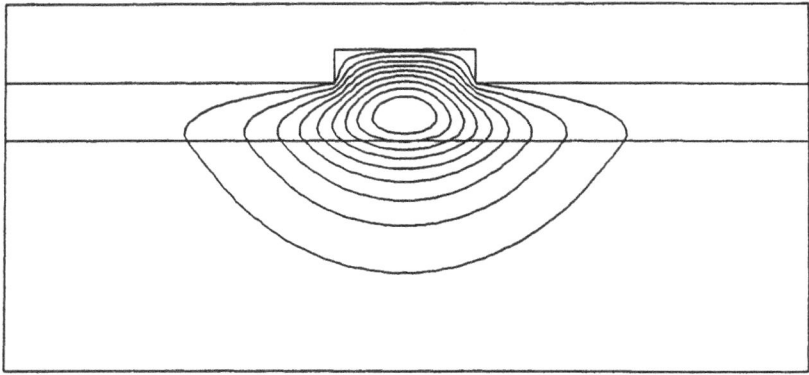

Figure 3.9: Contour plots of NFP for the E_{11}^x mode of rib waveguide structure 5.

Table 3.5: FWHMs for the E_{11}^x mode of rib waveguide structure 5 and for the E_{21}^x mode of rib waveguide structure 6.

Structure	Mode	Direction	EIM	SIM	SFEM
5	E_{11}^x	Horizontal	5.4°	4.5°	4.5°
		Vertical	11.1°	8.0°	8.3°
6	E_{21}^x	Horizontal	12.0°	11.4°	12.3°
		Peak position	$\theta_x = \pm 8.8°$	$\theta_x = \pm 6.3°$	$\theta_x = \pm 7.0°$

3.6.2 Rib waveguide junctions

Waveguide discontinuities play an important role in designing practical devices such as an isolated step discontinuity as in butt joints or as finite cascades such as gratings, tapers, bendings, or Y-junctions.

Considering an abrupt junction between two single-mode waveguides and using the effective index n_{eff}, the modal field $\phi(x, y)$, and the index profile $n(x, y)$ for each uniform waveguide, the normalized reflected power ξ and the normalized transmitted power η (coupling efficiency) are expressed approximately as follows:

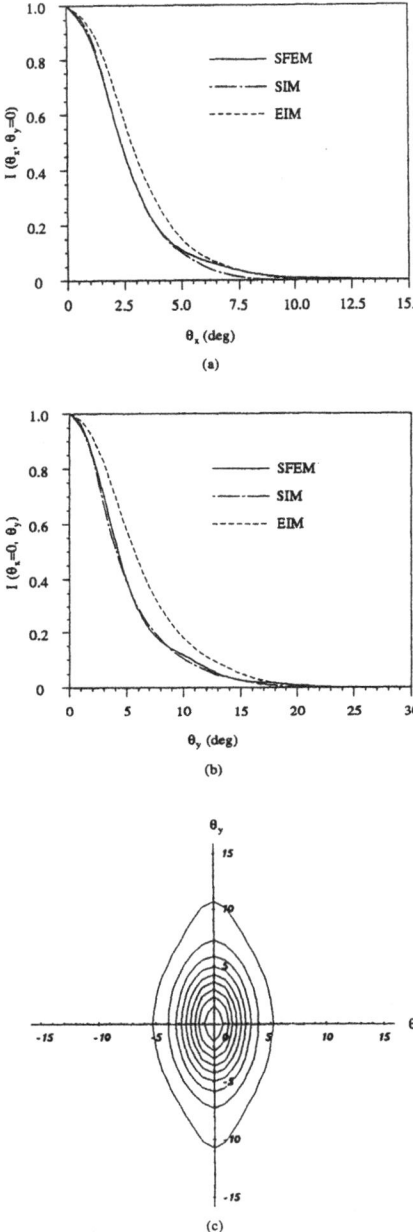

Figure 3.10: FFP for the E_{11}^x mode of rib waveguide structure 5. (a) Horizontal FFP. (b) Vertical FFP. (c) Contour plots of FFP.

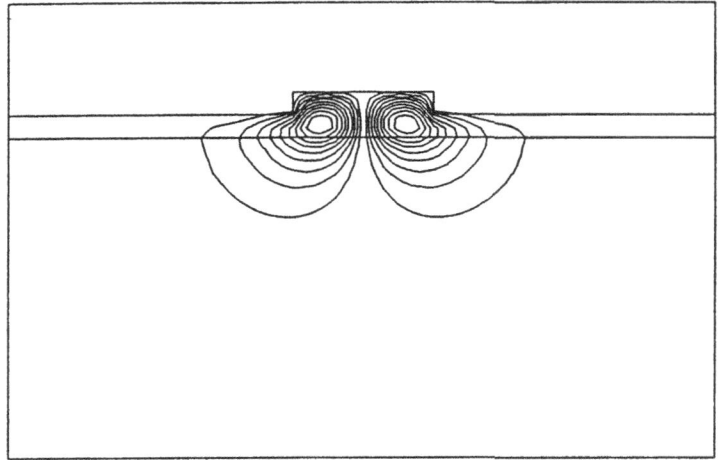

Figure 3.11: Contour plots of NFP for the E_{21}^x mode of rib waveguide structure 6.

(1) E^x modes:

$$\xi = \left(\frac{n_{\mathrm{eff1}} - n_{\mathrm{eff2}}}{n_{\mathrm{eff1}} + n_{\mathrm{eff2}}}\right)^2 \tag{3.87}$$

$$\eta = \frac{4 n_{\mathrm{eff1}} n_{\mathrm{eff2}} \left(\iint_{\Omega} \phi_1 \phi_2 \, dx \, dy\right)^2}{(n_{\mathrm{eff1}} + n_{\mathrm{eff2}})^2 \iint_{\Omega} \phi_1^2 \, dx \, dy \iint_{\Omega} \phi_2^2 \, dx \, dy} \tag{3.88}$$

where the subscripts 1 and 2 designate the quantities in input and output waveguides, respectively, and ϕ corresponds to E_x (see Eq. (3.74)).

(2) E^y modes:

$$\xi = \frac{\left(n_{\mathrm{eff1}} \iint_{\Omega} \frac{\phi_1 \phi_2}{n_1^2} \, dx \, dy - n_{\mathrm{eff2}} \iint_{\Omega} \frac{\phi_1 \phi_2}{n_2^2} \, dx \, dy\right)^2}{\left(n_{\mathrm{eff1}} \iint_{\Omega} \frac{\phi_1 \phi_2}{n_1^2} \, dx \, dy + n_{\mathrm{eff2}} \iint_{\Omega} \frac{\phi_1 \phi_2}{n_2^2} \, dx \, dy\right)^2} \tag{3.89}$$

$$\eta = \frac{4 n_{\mathrm{eff1}} n_{\mathrm{eff2}}}{\left(n_{\mathrm{eff1}} \iint_{\Omega} \frac{\phi_1 \phi_2}{n_1^2} \, dx \, dy + n_{\mathrm{eff2}} \iint_{\Omega} \frac{\phi_1 \phi_2}{n_2^2} \, dx \, dy\right)^2}$$

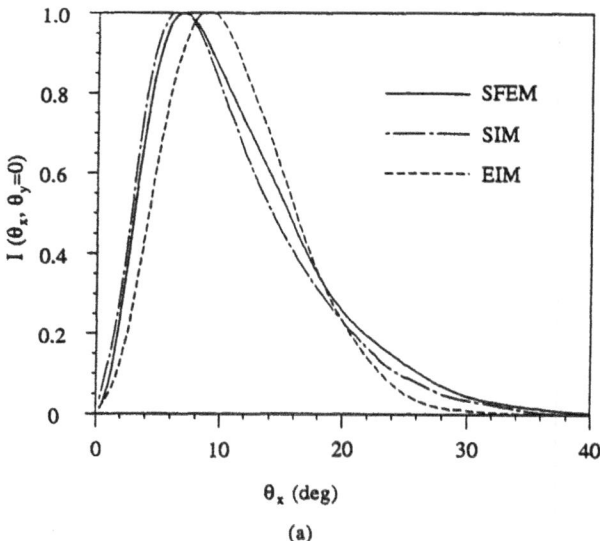

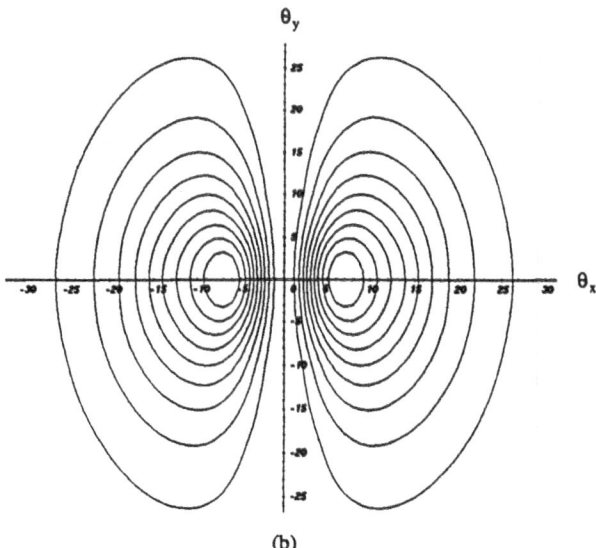

Figure 3.12: FFP for the E_{21}^x mode of rib waveguide structure 6. (a) Horizontal FFP. (b) Contour plots of FFP.

$$\times \frac{\left(\iint_{\Omega} \frac{\phi_1 \phi_2}{n_1^2}\, dx\, dy\right)^2 \left(\iint_{\Omega} \frac{\phi_1 \phi_2}{n_2^2}\, dx\, dy\right)^2}{\iint_{\Omega} \frac{\phi_1^2}{n_1^2}\, dx\, dy \iint_{\Omega} \frac{\phi_2^2}{n_2^2}\, dx\, dy} \qquad (3.90)$$

where ϕ corresponds to H_x (see Eq. (3.77)).

The effective index n_{eff} and the modal field $\phi(x,y)$ necessary to calculate the reflected and transmitted powers are generated by using the SFEM. The radiated power ζ is given by

$$\zeta = 1 - \xi - \eta. \qquad (3.91)$$

Figure 3.13 shows the coupling efficiency for the E_{11}^x and E_{11}^y modes of a rib waveguide junction based on structure 1 indicated in Table 3.2. The results of the very simple approach described here agree well with the more accurate ones calculated from a combination of the VFEM and the least square boundary residual method.[44]

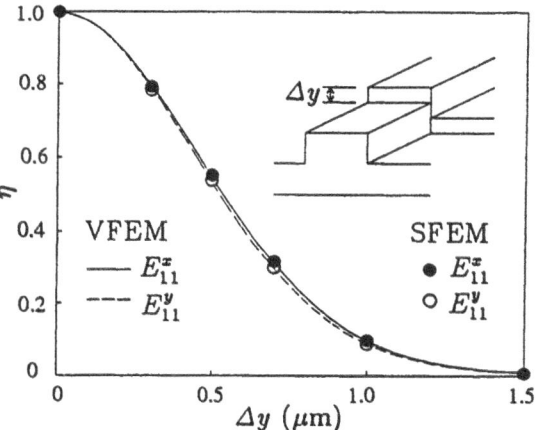

Figure 3.13: Coupling efficiency for the E_{11}^x and E_{11}^y modes of a rib waveguide with vertical displacement.

3.6.3 Rib waveguide directional couplers

Directional couplers, formed from two closely coupled waveguides, are important means of transferring a signal from one waveguide to an adjacent waveguide and are used for many applications in guided optics: for switching, power division, modulation, and frequency or polarization selection. Light launched into one waveguide (waveguide 1) will transfer to

the other waveguide (waveguide 2) after a characteristic distance known as
the coupling length. It will then pass back to the original waveguide, this
transfer of power occurring periodically down the length of the coupler.
From the coupled-mode theory the power flow in the two waveguides 1, 2
is expressed as

$$\eta_1 = 1 - F\sin^2\beta_b z \qquad (3.92a)$$

$$\eta_2 = F\sin^2\beta_b z \qquad (3.92b)$$

with

$$F = 1 - \left(\frac{\beta_1 - \beta_2}{\beta_e - \beta_o}\right)^2 = \frac{1}{1 + [(\beta_1 - \beta_2)/2\kappa]^2} \qquad (3.93)$$

$$\beta_b = \frac{\beta_e - \beta_o}{2} = \sqrt{\left(\frac{\beta_1 - \beta_2}{2}\right)^2 + \kappa^2} \qquad (3.94)$$

where β_e and β_o are the propagation constants for the lowest-order symmetric (even) and antisymmetric (odd) supermodes of the overall system,
and β_b, F, and κ are called the beat wavenumber, the power-conversion efficiency, and the coupling coefficient, respectively. If the coupler is asymmetric, then the propagation constants, β_1 and β_2, of the isolated waveguides
will not be the same. Complete power transfer will not be achieved, and
appreciable coupling will only occur if $|\beta_1 - \beta_2|$ is small. From Eqs. (3.92)
and (3.94) the coupling length L_c is given by

$$L_c = \pi/(\beta_e - \beta_o). \qquad (3.95)$$

Now, we consider directional couplers formed from the waveguides with
structures 1, 2, 3 indicated in Table 3.2 and separated by $S = 2\mu m$, and
subdivide only one-half of the waveguide cross section into quadratic triangular elements as shown in Fig. 3.14.

   Table 3.6 shows the normalized propagation constants b_e and b_o corresponding to β_e and β_o, respectively, and the coupling length L_c, where 256
elements and 537 nodes are used for structures 1, 2, and 268 elements and
567 nodes for structure 3. The SFEM values are seen to lie roughly within
the spread of values determined by the EIM,[98] ENM,[99] BPM,[100] and
SVM,[98] for the same waveguide structure. From the lack of consistency
in Table 3.6 one can conclude that accurate modeling is more difficult by
any method for rib waveguide couplers than it is for single rib waveguides
and that further research is necessary to assess the accuracy of the various
methods reported so far.

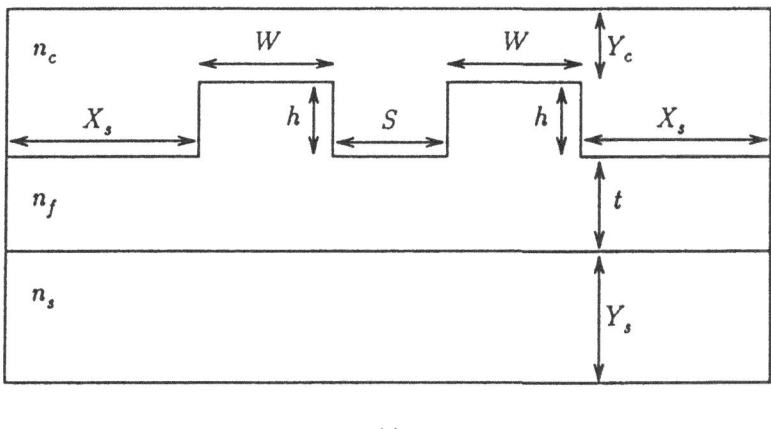

(a)

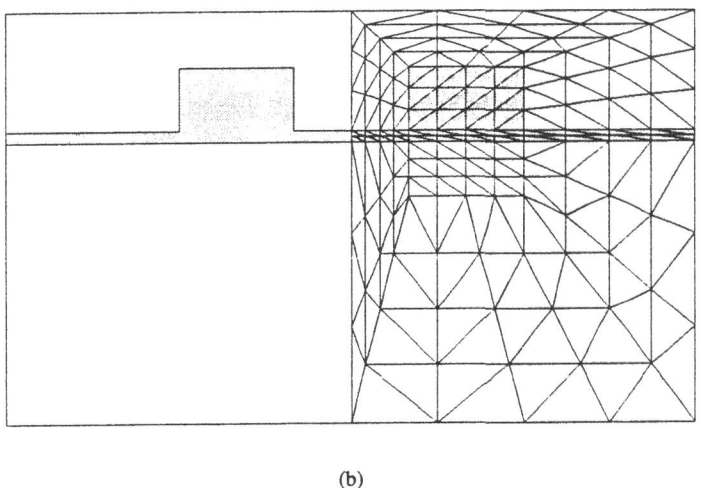

(b)

Figure 3.14: Rib waveguide directional coupler. (a) Waveguide structure. (b) Element division.

Table 3.6: Normalized propagation constants for the even and odd modes of directional couplers formed from the waveguides with structures 1, 2, 3 and separated by 2 μm, and the coupling length.

	Structure	EIM	ENM	BPM	SVM	SFEM
b_e	1	0.49953708	0.4782	0.518		0.50125762
	2	0.446	0.4436	0.459	0.436	0.44230827
	3	0.532	0.402	0.384	0.352	0.41821391
b_o	1	0.49953705	0.4782	0.506		0.50123658
	2	0.434	0.4319	0.446	0.424	0.43127565
	3	0.398		0.218	0.273	0.32988635
L_c (mm)	1	3×10^5		65.1		368.3
	2	0.807	0.818	0.71	0.827	0.8768
	3	1.152		0.93	1.968	1.755

References

References for the FEM using axial components of E and H fields

1) S. Ahmed and P. Daly, "Finite-element method for inhomogeneous waveguides", *IEE Proc.*, Vol. 116, No. 10, pp. 1661–1664, Oct. 1969.

2) P. Daly, "Finite-element coupling matrix", *Electron. Lett.*, Vol. 5, No. 24, pp 613–615, Nov. 1969.

3) Z. J. Csendes and P. Silvester, "Numerical solution of dielectric loaded waveguides I-Finite-element analysis", *IEEE Trans. Microwave Theory Tech.*, Vol. MTT-18 No. 12, pp. 1124–1131, Dec. 1970.

4) P. Daly, "Hybrid-mode analysis of microstrip by finite-element method", *IEEE Trans. Microwave Theory Tech.*, Vol. MTT-19, No. 1, pp. 19–25, Jan. 1971.

5) C. Yeh, S. B. Dong, and W. Oliver, "Arbitrarily shaped inhomogeneous optical fiber or integrated optical waveguides", *Jour. Appl. Phys.*, Vol. 46, No. 5, pp 2125–2129, May 1975.

6) P. Vandenbulcke and P. E. Lagasse, "Eigenmode analysis of anisotropic optical fibres or integrated optical waveguides", *Electron. Lett.*, Vol. 12, No. 5, pp 120–122, March 1976.

7) A. D. McAulay, "The finite element solution of dissipative electromagnetic surface waveguides", *Int. Jour. Numer. Methods Eng.*, Vol. 11, pp. 11–25, 1977.

8) A. D. McAulay, "Variational finite-element solution for dissipative waveguides and transportation application", *IEEE Trans. Microwave Theory Tech.*, Vol. MTT-25, No. 5, pp. 382–392, May 1977.

9) T. S. Bird, "Propagation and radiation characteristics of rib waveguide", *Electron. Lett.*, Vol. 13, No. 14, pp. 401–403, July 1977.

10) P. M. Pelosi, P. Vandenbulcke, C. D. W. Wilkinson, and R. M. De La Rue, "Propagation characteristics of trapezoidal cross-section ridge optical waveguides", *Electron. Lett.*, Vol. 13, No. 20, pp. 607–608, Sept. 1977.

11) P. M. Pelosi, P. Vandenbulcke, C. D. W. Wilkinson, and R. M. De La Rue, "Propagation characteristics of trapezoidal cross-section ridge optical waveguides: An experimental and theoretical investigation", *Appl. Opt.*, Vol. 17, No. 8, pp. 1187–1193, April 1978.

12) C. Yeh, K. Ha, S. B. Dong, and W. P. Brown, "Single-mode optical waveguides", *Appl. Opt.*, Vol. 18, No. 10, pp. 1490–1504, May 1979.

13) M. Ikeuchi, H. Sawami, and H. Niki, "Analysis of open-type dielectric waveguides by the finite-element iterative method", *IEEE Trans. Microwave Theory Tech.*, Vol. MTT-29, No. 3, pp. 234–239, March 1981.

14) N. Mabaya, P. E. Lagasse, and P. Vandenbulcke, "Finite element analysis of optical waveguides", *IEEE Trans. Microwave Theory Tech.*, Vol. MTT-29, No. 6, pp. 600–605, June 1981.

15) M. Aubourg, J. P. Villotte, F. Godon, and Y. Garault, "Finite element analysis of lossy waveguides—Application to microstrip lines on semiconductor substrate", *IEEE Trans. Microwave Theory Tech.*, Vol. MTT-31, No. 4, pp. 326–331, April 1983.

16) R. B. Wu and C. H. Chen, "On the variational reaction theory for dielectric waveguides", *IEEE Trans. Microwave Theory Tech.*, Vol. MTT-33, No. 6, pp. 477–483, June 1985.

17) D. Welt and J. Webb, "Finite-element analysis of dielectric waveguides with curved boundaries", *IEEE Trans. Microwave Theory Tech.*, Vol. MTT-33, No. 7, pp. 576–586, July 1985.

18) R. B. Wu and C. H. Chen, "A variational analysis of dielectric waveguides by the conformal mapping technique", *IEEE Trans. Microwave Theory Tech.*, Vol. MTT-33, No. 8, pp. 681–685, Aug. 1985.

19) C. K. Tzuang and T. Itoh, "Finite-element analysis of slow-wave Schottky contact printed lines", *IEEE Trans. Microwave Theory Tech.*, Vol. MTT-34, No. 12, pp. 1483–1489, Dec. 1986.

20) Eswarappa, G. I. Costache, and W. J. R. Hoefer, "Finlines in rectangular and circular waveguide housings including substrate mounting and bending effects— Finite element analysis", *IEEE Trans. Microwave Theory Tech.*, Vol. 37, No. 2, pp. 299–306, Feb. 1989.

21) A. A. P. Gibson and J. Helszajn, "Finite element solution of longitudinally magnetized elliptical gyromagnetic waveguides", *IEEE Trans. Microwave Theory Tech.*, Vol. 37, No. 6, pp. 999–1005, June 1989.

22) C. N. Chang, Y. C. Wong, and C. H. Chen, "Full-wave analysis of coplanar waveguides by variational conformal mapping technique", *IEEE Trans. Microwave Theory Tech.*, Vol. 38, No. 9, pp. 1339–1344, Sept. 1990.

References for the FEM using a full vector E or H field

23) A. Konrad, "High-order triangular finite elements for electromagnetic waves in anisotropic media", *IEEE Trans. Microwave Theory Tech.*, Vol. MTT-25, No. 5, pp. 353–360, May 1977.

24) B. M. A. Rahman and J. B. Davies, "Finite-element analysis of optical and microwave waveguide problems", *IEEE Trans. Microwave Theory Tech.*, Vol. MTT-32, No. 1, pp. 20–28, Jan. 1984.

25) M. Koshiba, K. Hayata, and M. Suzuki, "Vectorial finite-element formulation without spurious modes for dielectric waveguides", *Trans. Inst. Electron. Commun. Eng. Japan*, Vol. E67, No. 4, pp. 191–196, April 1984.

26) M. Koshiba, K. Hayata, and M. Suzuki, "Vectorial finite-element formulation without spurious solutions for dielectric waveguide problems", *Electron. Lett.*, Vol. 20, No. 10, pp. 409–410, May 1984.

27) B. M. A. Rahman and J. B. Davies, "Penalty function improvement of waveguide solution by finite elements", *IEEE Trans. Microwave Theory Tech.*, Vol. MTT-32, No. 8, pp. 922–928, Aug. 1984.

28) J. B. Davies and B. M. A. Rahman, "Analysis of open optical and microwave guides of arbitrary transverse permittivity profiles", *Radio Sci.*, Vol. 19, No. 5, pp. 1245–1249, Sept.–Oct. 1984.

29) M. Hano, "Finite-element analysis of dielectric-loaded waveguides", *IEEE Trans. Microwave Theory Tech.*, Vol. MTT-32, No. 10, pp. 1275–1279, Oct. 1984.

30) B. M. A. Rahman and J. B. Davies, "Finite-element solution of integrated optical waveguides", *IEEE/OSA Jour. Lightwave Technol.*, Vol. LT-2, No. 5, pp. 682–688, Oct. 1984.

31) M. Koshiba, K. Hayata, and M. Suzuki, "Improved finite-element formulation in terms of magnetic field vector for dielectric waveguides", *IEEE Trans. Microwave Theory Tech.*, Vol. MTT-33, No. 3, pp. 227–233, March 1985.

32) M. Koshiba, K. Hayata, and M. Suzuki, "Finite-element formulation in terms of the electric-field vector for electromagnetic waveguide problems", *IEEE Trans. Microwave Theory Tech.*, Vol. MTT-33, No. 10, pp. 900–905, Oct. 1985.

33) B. M. A. Rahman and J. B. Davies, "Vector-H finite element solution of GaAs/GaAlAs rib waveguides", *IEE Proc.*, Vol. 132, Pt. J, No. 6, pp. 349–353, Dec. 1985.

34) M. Koshiba, K. Hayata, and M. Suzuki, "Vector-E field finite-element analysis of dielectric optical waveguides", *Appl. Opt.*, Vol. 25, No. 1, pp. 10–11, Jan. 1986.

35) M. Koshiba, K. Hayata, and M. Suzuki, "Finite-element solution of anisotropic waveguides with arbitrary tensor permittivity", *IEEE/OSA Jour. Lightwave Technol.*, Vol. LT-4, No. 2, pp. 121–126, Feb. 1986.

36) K. Hayata, M. Koshiba, M. Eguchi, and M. Suzuki, "Novel finite-element formulation without any spurious solutions for dielectric waveguides", *Electron. Lett.*, Vol. 22, No. 6, pp. 295–296, March 1986.

37) A. J. Kobelansky and J. B. Webb, "Eliminating spurious modes in finite-element waveguide problems by using divergence-free fields", *Electron. Lett.*, Vol. 22, No. 11, pp. 569–570, May 1986.

38) K. Hayata, M. Koshiba, and M. Suzuki, "Lateral mode analysis of buried heterostructure diode lasers by the finite-element method", *IEEE Jour. Quantum Electron.*, Vol. QE-22, No. 6, pp. 781–788, June 1986.

39) T. P. Young and P. Smith, "Finite element modelling of integrated optical waveguides", *GEC Jour. Res.*, Vol. 4, pp. 249–255, 1986.

40) K. Hayata, M. Koshiba, M. Eguchi, and M. Suzuki, "Vectorial finite-element method without any spurious solutions for dielectric waveguiding problem using transverse magnetic-field component", *IEEE Trans. Microwave Theory Tech.*, Vol. MTT-34, No. 11, pp. 1120–1124, Nov. 1986.

41) F. Kikuchi, "Mixed and penalty formulations for finite element analysis of an eigenvalue problem in electromagnetism", *Comput. Methods Appl. Mech. Eng.*, Vol. 64, pp. 509–521, 1987.

42) K. Hayata, K. Miura, and M. Koshiba, "Finite-element formulation for lossy waveguides", *IEEE Trans. Microwave Theory Tech.*, Vol. 36, No. 2, pp. 268–276, Feb. 1988.

43) T. P. Young, "Design of integrated optical circuits using finite elements", *IEE Proc.*, Vol. 135, Pt. A, No. 3, pp. 135–144, March 1988.

44) B. M. A. Rahman and J. B. Davies, "Analysis of optical waveguide discontinuities", *IEEE/OSA Jour. Lightwave Technol.*, Vol. 6, No. 1, pp. 52–57, Jan. 1988.

45) B. M. A. Rahman and J. B. Davies, "Analysis of optical waveguides and some discontinuity problems", *IEE Proc.*, Vol. 135, Pt. J, No. 5, pp. 339–342, Oct. 1988.

46) S. H. Wong and Z. J. Cendes, "Combined finite element-modal solution of three-dimensional eddy current problems", *IEEE Trans. Magnet.*, Vol. 24, No. 6, pp. 2685–2687, Nov. 1988.

47) J. P. Webb, "Finite element analysis of dispersion in waveguides with sharp metal edges", *IEEE Trans. Microwave Theory Tech.*, Vol. 36, No. 12, pp. 1819–1824, Dec. 1988.

48) K. Hayata, M. Eguchi, and M. Koshiba, "Finite element formulation for guided-wave problems using transverse electric field component", *IEEE Trans. Microwave Theory Tech.*, Vol. 37, No. 1, pp. 256–258, Jan. 1989.

49) K. Hayata, K. Miura, and M. Koshiba, "Full vectorical finite element formulation for lossy anisotropic waveguides", *IEEE Trans. Microwave Theory Tech.*, Vol. 37, No. 5, pp. 875–883, May 1989.

50) G. W. Slade and K. J. Webb, "A vectorial finite element analysis for integrated waveguide", *IEEE Trans. Magnet.*, Vol. 25, No. 4, pp. 3052–3054, July 1989.

51) H. Yonezawa and K. Sakuda, "Numerical analysis of optically coupled active and passive dielectric slab waveguides for TM modes", *Appl. Opt.*, Vol. 28, No. 17, pp. 3581–3594, Sept. 1989.

52) C. De Bernaridi, S. Morasca, C. Rigo, B. Sordo, A. Stano, I. R. Croston, and T. P. Young, "Wavelength demultiplexer integrated on AlGaInAs/InP for 1.5μm operation", *Electron. Lett.*, Vol. 25, No. 22, pp. 1488–1489, Oct. 1989.

53) M. J. McDougall and J. P. Webb, "Infinite elements for the analysis of open dielectric waveguides", *IEEE Trans. Microwave Theory Tech.*, Vol. 37, No. 11, pp. 1724–1731, Nov. 1989.

54) M. Israel and R. Miniowitz, "Hermitian finite-element method to inhomogeneous waveguides", *IEEE Trans. Microwave Theory Tech.*, Vol. 38, No. 9, pp. 1319–1327, Sept. 1990.

55) K. Hayakawa and K. Sakuda, "Comparison of numerical simulation on active rectangular waveguide couplers by vector, scalar finite element, and effective index methods", *Appl. Opt.*, Vol. 29, No. 21, pp. 3100–3109, July 1990.

56) K. D. Paulsen and D. R. Lynch, "Elimination of parasites in finite element Maxwell solutions", *IEEE Trans. Microwave Theory Tech.*, Vol. 39, No. 3, pp. 395–404, March 1991.

57) R. Miniowitz and J. P. Webb, "Covariant-projection quadrilateral elements for the analysis of waveguides with sharp edges", *IEEE Trans. Microwave Theory Tech.*, Vol. 39, No. 3, pp. 501–505, March 1991.

58) J. F. Lee, D. K. Sun, and Z. J. Cendes, "Full-wave analysis of dielectric waveguides using tangential vector finite elements", *IEEE Trans. Microwave Theory Tech.*, Vol. 39, No. 8, pp. 1262–1271, Aug. 1991.

59) E. T. Moyer and E. A. Schroeder, "Finite element formulations of Maxwell's equations — Advantages and comparisons between available approaches", *IEEE Trans. Magnet.*, Vol. 27, No. 5, pp. 4217–4220, Sept. 1991.

60) M. Matsuhara, H. Yunoki, and A. Maruta, "Analysis of open-type waveguides by the vector finite-element method", *IEEE Microwave Guided Wave Lett.*, Vol. 1, No. 12, pp. 376–378, Dec. 1991.

61) G. W. Slade and K. J. Webb, "Computation of characteristic impedance for multiple microstrip transmission lines using a vector finite element method", *IEEE Trans. Microwave Theory Tech.*, Vol. 40, No. 1, pp. 34–40, Jan. 1992.

62) M. Koshiba and K. Inoue, "Simple and efficient finite-element analysis of microwave and optical waveguides", *IEEE Trans. Microwave Theory Tech.*, Vol. 40, No. 2, pp. 371–377, Feb. 1992.

References for the FEM using both E and H fields

63) J. A. M. Svedin, "A numerically efficient finite-element formulation for the general waveguide problem without spurious modes", *IEEE Trans. Microwave Theory Tech.*, Vol. 37, No. 11, pp. 1708–1715, Nov. 1989.

64) J. A. M. Svedin, "Finite-element analysis of chirowaveguides", *Electron. Lett.*, Vol. 26, No. 13, pp. 928–929, June 1990.

65) J. A. M. Svedin, "Propagation analysis of chirowaveguides using finite-element method", *IEEE Trans. Microwave Theory Tech.*, Vol. 38, No. 10, pp. 1488–1496, Oct. 1990.

66) J. A. M. Svedin, "A modified finite-element method for dielectric waveguides using an asymptotically correct approximation on infinite elements", *IEEE Trans. Microwave Theory Tech.*, Vol. 39, No. 2, pp. 258–266, Feb. 1991.

References for the FEM using transverse components of an *E* or *H* field

67) J. Katz, "Novel solution of 2-D waveguides using the finite element method", *Appl. Opt.*, Vol. 21, No. 15, pp. 2747–2750, Aug. 1982.

68) C. C. Su, "A combined method for dielectric waveguides using the finite-element technique and the surface integral equations method", *IEEE Trans. Microwave Theory Tech.*, Vol. MTT-34, No. 11, pp. 1140–1146, Nov. 1986.

69) Z. P. Tanner, C. H. Chan, and R. Mittra, "Finite-element analysis of anisotropic waveguides with storage reduction and elimination of spurious modes", *Microwave Opt. Technol. Lett.*, Vol. 2, No. 1, pp. 3–6, Jan. 1989.

70) W. C. Chew and M. A. Nasir, "A variational analysis of anisotropic, inhomogeneous dielectric waveguides", *IEEE Trans. Microwave Theory Tech.*, Vol. 37, No. 4, pp 661–668, April 1989.

71) F. A. Fernandez and Y. Lu, "Variational finite element analysis of dielectric waveguides with no spurious solutions", *Electron. Lett.*, Vol. 26, No. 25, pp. 2125–2126, Dec. 1990.

72) F. A. Fernandez, J. B. Davies, S. Zhu, and Y. Lu, "Sparse matrix eigenvalue solver for finite element solution of dielectric waveguides", *Electron. Lett.*, Vol. 27, No. 20, pp. 1824–1826, Sept. 1991.

73) F. A. Fernandez and Y. Lu, "A variational finite element formulation for dielectric waveguides in terms of transverse magnetic fields", *IEEE Trans. Magnet.*, Vol. 27, No. 5, pp. 3864–3867, Sept. 1991.

References for the FEM using transverse components of both *E* and *H* fields

74) T. Angkaew, M. Matsuhara, and N. Kumagai, "Finite-element analysis of waveguide modes: A novel approach that eliminates spurious modes", *IEEE Trans. Microwave Theory Tech.*, Vol. MTT-35, No. 2, pp. 117–123, Feb. 1987.

75) T. Angkaew, M. Matsuhara, and N. Kumagai, "An improved finite-element formulation using transverse electric or magnetic fields components for electromagnetic waveguide mode analysis", *Trans. Inst. Electron. Inform. Commun. Eng.*, Vol. E70, No. 9, pp. 841–846, Sept. 1987.

76) C. Forterre, P. H. Giesbers, and E. Laroche, "Finite element analysis of ferrite loaded transmission lines", *IEEE Trans. Magnet.*, Vol. MAG-23, No. 5, pp. 2666–2667, Sept. 1987.

77) M. Matsumoto, "Analysis of leakage properties of periodic dielectric image guides", *Electron. Lett.*, Vol. 25, No. 23, pp. 1568–1569, Nov. 1989.

78) M. Matsumoto, "Analysis of radiation properties of channel-waveguide grating", *Jour. Opt. Soc. Am. B*, Vol. 8, No. 2, pp. 434–442, Feb. 1991.

79) M. Auboung and P. Guillon, "A mixed finite element formulation for microwave devices problems. Application to MIS structure", *Jour. Electromagnet. Waves Applic.*, Vol. 5, No. 4/5, pp. 371–386, 1991.

Reference for the FEM using vector and scalar potentials

80) I. Bárdi and O. Bíró, "Improved finite element formulation for dielectric loaded waveguides", *IEEE Trans. Magnet.*, Vol. 26, No. 2, pp. 450–453, March 1990.

81) I. Bárdi and O. Bíró, "An efficient finite-element formulation without spurious modes for anisotropic waveguides", *IEEE Trans. Microwave Theory Tech.*, Vol. 39, No. 7, pp. 1133–1139, July 1991.

References for the approximate scalar FEM

82) M. Koshiba, K. Hayata, and M. Suzuki, "Approximate scalar finite-element analysis of anisotropic optical waveguides", *Electron. Lett.*, Vol. 18, No. 10, pp. 411–413, May 1982.

83) M. Koshiba, K. Hayata, and M. Suzuki, "On accuracy of approximate scalar finite-element analysis of dielectric optical waveguides", *Trans. Inst. Electron. Commun. Eng. Japan* , Vol. E66, No. 2, pp. 157–158, Feb. 1983.

84) M. Koshiba, K. Hayata, and M. Suzuki, "Approximate scalar finite-element analysis of anisotropic optical waveguides with off-diagonal elements in a permittivity tensor", *IEEE Trans. Microwave Theory Tech.*, Vol. MTT-32, No. 6, pp. 587–593, June 1984.

85) K. S. Chiang, "Finite element method for cutoff frequencies of weakly guiding fibres of arbitrary cross-section", *Opt. Quantum Electron.*, Vol. 16, pp. 487–493, 1984.

86) K. S. Chiang, "Finite-element analysis of optical fibres with iterative treatment of the infinite 2-D space", *Opt. Quantum Electron.*, Vol. 17, pp. 381–391, 1985.

87) R. B. Wu and C. H. Chen, "A scalar variational conformal mapping technique for weakly guiding dielectric waveguides", *IEEE Jour. Quantum Electron.*, Vol. QE-22, No. 5, pp. 603–609, May 1986.

88) K. S. Chiang, "Finite element analysis of weakly guiding fibers with arbitrary refractive-index distribution", *IEEE/OSA Jour. Lightwave Technol.*, Vol. LT-4, No. 8, pp. 980–990, Aug. 1986.

89) K. Hayata, M. Eguchi, and M. Koshiba, "Self-consistent finite/infinite element for unbounded guided wave problems", *IEEE Trans. Microwave Theory Tech.*, Vol. 36, No. 3, pp. 614–616, March 1988.

90) E. Strake, G. P. Bava, and I. Montrosset, "Guided modes of Ti:LiNbO₃ channel waveguides: A novel quasi-analytical technique in comparison with the scalar finite-element method", *IEEE/OSA Jour. Lightwave Technol.*, Vol. 6, No. 6, pp. 1126–1135, June 1988.

91) R. V. Mustacich, "Scalar finite element analysis of electrooptic modulation in diffused channel waveguides and poled waveguides in polymer thin films", *Appl. Opt.*, Vol. 27, No. 17, pp. 3732–3737, Sept. 1988.

92) C. Neubauer, R. März, and M. Shienle, "A comparison between finite element calculations and experimental results on InGaAsP/ InP waveguides", *IEEE/OSA Jour. Lightwave Technol.*, Vol. 8, No. 12, pp. 1932–1936, Dec. 1990.

93) J. C. Grant, J. C. Beal, and N. J. P. Frenette, "Solving certain leaky waveguides with lossless, simply bounded finite element modeling", *IEEE Photonics Technol. Lett.*, Vol. 2, No. 12, pp. 890-892, Dec. 1990.

94) L. Bersiner, U. Hempelmann, and E. Strake, "Numerical analysis of passive integrated-optical polarization splitters: comparison of finite-element method and beam-propagation method results", *Jour. Opt. Soc. Am. B* , Vol. 8, No. 2, pp. 422-433, Feb. 1991.

95) M. Koshiba, H. Saitoh, M. Eguchi, and K. Hirayama, "A simple scalar finite element approach to optical rib waveguides", *IEE Proc.*, Vol. 139, Pt. J, No. 2, pp. 166–171, April 1992.

References for the other optical waveguide theories

96) E. A. J. Marcatili, "Dielectric rectangular waveguide and directional coupler for integrated optics", *Bell Syst. Tech. Jour.*, Vol. 48, No. 7, pp. 2071–2102, Sept. 1969.

97) J. E. Goell, "A circular harmonic computer analysis of rectangular dielectric waveguides", *Bell Syst. Tech. Jour.*, Vol. 48, No. 7, pp. 2133–2160, Sept. 1969.

98) M. J. Robertson, S. Ritchie, and P. Dayan, "Semiconductor waveguide: Analysis of optical propagation in single rib structures and directional couplers", *IEE Proc.*, Vol. 132, Pt. J, No. 6, pp. 336–342, Dec. 1985.

99) N. Dagli and C. G. Fonstand, "Theoretical and experimental study of the analysis and modeling of integrated optical components", *IEEE Jour. Quantum Electron.*, Vol. 24, No. 11, pp. 2215–2226, Nov. 1988.

100) M. D. Feit and J. A. Fleck, Jr., "Analysis of rib waveguides and couplers by the propagating beam method", *Jour. Opt. Soc. Am. A*, Vol. 7, No. 1, pp. 73–79 Jan. 1990.

101) M. S. Stern, "Semivectorial polarized finite difference method for optical waveguides with arbitrary index profiles", *IEE Proc.*, Vol. 135, Pt. J, No. 1, pp. 56–63, Feb. 1988.

102) P. N. Robson and P. C. Kendall (eds.), *Rib Waveguide Theory by the Spectral Index Method*, Research Studies Press, Taunton, 1990.

Chapter 4

OPTICAL FIBERS

4.1 Introduction

Single-mode optical fibers have already been one of the major transmission media for long-distance telecommunication, with very low loss and very high bandwidth.

For the finite element analysis of the hybrid modes of optical fibers the variational expression in terms of axial components of the electromagnetic field is generally utilized.[1),2)] In applying this approach, great attention should be paid to how to assign nodes.[2)] We write the refractive indices at the neighboring nodes i and j as n_i and n_j, respectively. When the effective index n_{eff} approaches $n_{\text{eff}} \simeq \sqrt{(n_i^2 + n_j^2)/2}$, the matrix derived from discretization is numerically ill-conditioned, and consequently spurious solutions can be involved.[2)] If we want to avoid such solutions, it is necessary to assign the node i to the position $r = r_i$ where the refractive index is equal to the effective index; that is, $n(r_i) = n_{\text{eff}}$.[2)] However, because the term $1/[n^2(r_i) - n_{\text{eff}}^2]$ is involved in the variational expression in terms of axial electromagnetic field, an integrand included in the variational expression diverges at $r = r_i$.

In this chapter, to avoid such a difficulty, the FEM in terms of three components of the magnetic field is described for the vectorial wave analysis of optical fibers, and the approximate scalar FEM[3)−6)] is also introduced.

4.2 Method of Vectorial Wave Analysis

4.2.1 Basic equations

We consider an axially symmetrical optical fiber in Fig. 4.1, where $n(r)$ and n_2 are the refractive indices of the inhomogeneous core $(0 \leq r \leq a)$ and homogeneous cladding $(r \geq a)$ regions, respectively.

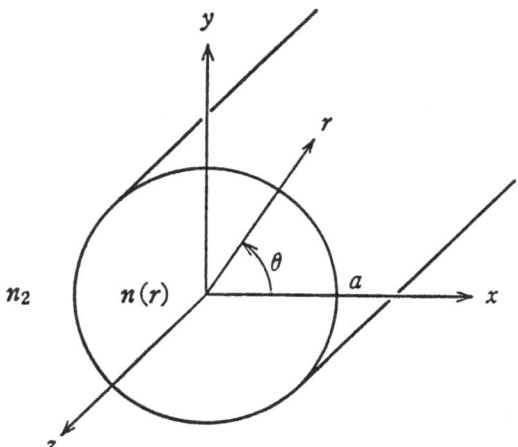

Figure 4.1: Optical fiber.

With a time dependence of the form $\exp(j\omega t)$ and a z-dependence of the form $\exp(-j\beta z)$ being implied, from Maxwell's equations the following vectorial wave equation is derived:

$$\nabla \times \left(\frac{1}{n^2}\nabla \times \boldsymbol{H}\right) - k_0^2 \boldsymbol{H} = 0\,. \tag{4.1}$$

Using the penalty function method (see Subsection 3.3.2), the functional for Eq. (4.1) is written as

$$
\begin{aligned}
F \;=\; & \iint_\Omega \left[(\nabla \times \boldsymbol{H})^* \cdot \left(\frac{1}{n^2}\nabla \times \boldsymbol{H}\right) - k_0^2 \boldsymbol{H}^* \cdot \boldsymbol{H}\right] r\,dr\,d\theta \\
& + s \iint_\Omega (\nabla \cdot \boldsymbol{H})^*(\nabla \cdot \boldsymbol{H})\,r\,dr\,d\theta \\
& + \int_\Gamma \boldsymbol{H}^* \cdot \left[\boldsymbol{i}_r \times \left(\frac{1}{n_2^2}\nabla \times \boldsymbol{H}\right)\right] d\Gamma
\end{aligned}
\tag{4.2}
$$

where Ω is the core region, Γ is the boundary between the core and cladding regions, and $\boldsymbol{i}_r$ is the unit vector in the r direction.

4.2.2 Finite element approach

Dividing the core region into a number of quadratic ring elements (see Fig. 1.5(b)), we expand the magnetic field $\boldsymbol{H}$ in each element as

$$\boldsymbol{H} = [N]^{\mathrm{T}}\{H\}_e \exp(-jm\theta) \tag{4.3}$$

with

$$\{H\}_e = \begin{bmatrix} \{H_r\}_e \\ \{H_\theta\}_e \\ \{H_z\}_e \end{bmatrix} \tag{4.4}$$

$$[N] = \begin{bmatrix} \{N\} & \{0\} & \{0\} \\ \{0\} & j\{N\} & \{0\} \\ \{0\} & \{0\} & j\{N\} \end{bmatrix} \tag{4.5}$$

where $\{H_r\}_e$, $\{H_\theta\}_e$, and $\{H_z\}_e$ are the nodal magnetic-field vectors.

Substituting Eq. (4.3) into Eq. (4.2), we obtain

$$\begin{aligned} F = & \{H\}^{\mathrm{T}}([K] + s[L] - k_0^2[M])\{H\} \\ & + \int_\Gamma \boldsymbol{H}^* \cdot \left[\boldsymbol{i}_r \times \left(\frac{1}{n_2^2} \nabla \times \boldsymbol{H} \right) \right] d\Gamma \end{aligned} \tag{4.6}$$

with

$$\begin{aligned} [K] &= \begin{bmatrix} [K_{rr}] & [K_{r\theta}] & [K_{rz}] \\ [K_{\theta r}] & [K_{\theta\theta}] & [K_{\theta z}] \\ [K_{zr}] & [K_{z\theta}] & [K_{zz}] \end{bmatrix} \\ &= 2\pi \sum_e \int_e \frac{1}{n^2(r)} [B]^* [B]^{\mathrm{T}} r \, dr \end{aligned} \tag{4.7}$$

$$\begin{aligned} [L] &= \begin{bmatrix} [L_{rr}] & [L_{r\theta}] & [L_{rz}] \\ [L_{\theta r}] & [L_{\theta\theta}] & [L_{\theta z}] \\ [L_{zr}] & [L_{z\theta}] & [L_{zz}] \end{bmatrix} \\ &= 2\pi \sum_e \int_e \{C\}\{C\}^{\mathrm{T}} r \, dr \end{aligned} \tag{4.8}$$

$$\begin{aligned} [M] &= \begin{bmatrix} [M_{rr}] & [0] & [0] \\ [0] & [M_{\theta\theta}] & [0] \\ [0] & [0] & [M_{zz}] \end{bmatrix} \\ &= 2\pi \sum_e \int_e [N]^* [N]^{\mathrm{T}} r \, dr \end{aligned} \tag{4.9}$$

$$[B] = \begin{bmatrix} \{0\} & -j\beta\{N\} & jm\{N\}/r \\ -\beta\{N\} & \{0\} & j\{N_r\} + j\{N\}/r \\ m\{N\}/r & -j\{N_r\} & \{0\} \end{bmatrix} \tag{4.10}$$

$$\{C\} = \begin{bmatrix} \{N_r\} + \{N\}/r \\ m\{N\}/r \\ \beta\{N\} \end{bmatrix} \tag{4.11}$$

where the submatrices of $[K]$, $[L]$, and $[M]$ are given by

$$[K_{rr}] = 2\pi \sum_e \int_e \frac{1}{n^2} \left[\beta^2 r\{N\}\{N\}^{\mathrm{T}} + \frac{m^2}{r}\{N\}\{N\}^{\mathrm{T}} \right] dr \qquad (4.12a)$$

$$[K_{r\theta}] = [K_{\theta r}]^{\mathrm{T}} = 2\pi \sum_e \int_e \frac{1}{n^2} \left[m\{N\}\{N_r\}^{\mathrm{T}} + \frac{m}{r}\{N\}\{N\}^{\mathrm{T}} \right] dr$$

$$\qquad (4.12b)$$

$$[K_{rz}] = [K_{zr}]^{\mathrm{T}} = 2\pi \sum_e \int_e \frac{1}{n^2} \beta r\{N\}\{N_r\}^{\mathrm{T}} \, dr \qquad (4.12c)$$

$$[K_{\theta\theta}] = 2\pi \sum_e \int_e \frac{1}{n^2} \left[\beta^2 r\{N\}\{N\}^{\mathrm{T}} + r\{N_r\}\{N_r\}^{\mathrm{T}} + \{N_r\}\{N\}^{\mathrm{T}} \right.$$

$$\left. + \{N\}\{N_r\}^{\mathrm{T}} + \frac{1}{r}\{N\}\{N\}^{\mathrm{T}} \right] dr \qquad (4.12d)$$

$$[K_{\theta z}] = [K_{z\theta}]^{\mathrm{T}} = 2\pi \sum_e \int_e -\frac{1}{n^2} m\beta\{N\}\{N\}^{\mathrm{T}} \, dr \qquad (4.12e)$$

$$[K_{zz}] = 2\pi \sum_e \int_e \frac{1}{n^2} \left[\frac{m^2}{r}\{N\}\{N\}^{\mathrm{T}} + r\{N_r\}\{N_r\}^{\mathrm{T}} \right] dr \qquad (4.12f)$$

$$[L_{rr}] = 2\pi \sum_e \int_e \left[r\{N_r\}\{N_r\}^{\mathrm{T}} + \{N_r\}\{N\}^{\mathrm{T}} \right.$$

$$\left. + \{N\}\{N_r\}^{\mathrm{T}} + \frac{1}{r}\{N\}\{N\}^{\mathrm{T}} \right] dr \qquad (4.13a)$$

$$[L_{r\theta}] = [L_{\theta r}]^{\mathrm{T}} = 2\pi \sum_e \int_e \left[m\{N_r\}\{N\}^{\mathrm{T}} + \frac{m}{r}\{N\}\{N\}^{\mathrm{T}} \right] dr$$

$$\qquad (4.13b)$$

$$[L_{rz}] = [L_{zr}]^{\mathrm{T}} = 2\pi \sum_e \int_e \left[\beta r\{N_r\}\{N\}^{\mathrm{T}} + \beta\{N\}\{N\}^{\mathrm{T}} \right] dr$$

$$\qquad (4.13c)$$

$$[L_{\theta\theta}] = 2\pi \sum_e \int_e \frac{m^2}{r}\{N\}\{N\}^{\mathrm{T}} \, dr \qquad (4.13d)$$

$$[L_{\theta z}] = [L_{z\theta}]^{\mathrm{T}} = 2\pi \sum_e \int_e m\beta\{N\}\{N\}^{\mathrm{T}} \, dr \qquad (4.13e)$$

$$[L_{zz}] = 2\pi \sum_e \int_e \beta^2 r\{N\}\{N\}^{\mathrm{T}} \, dr \qquad (4.13f)$$

$$[M_{rr}] = [M_{\theta\theta}] = [M_{zz}] = 2\pi \sum_e \int_e r\{N\}\{N\}^{\mathrm{T}} \, dr \, . \tag{4.14}$$

4.2.3 Analytical approach

The axial-field components, E_z and H_z, in the cladding region are given by

$$E_z = jAK_m \left(\frac{wr}{a}\right) \exp(-jm\theta) \tag{4.15}$$

$$H_z = BK_m \left(\frac{wr}{a}\right) \exp(-jm\theta) \tag{4.16}$$

with

$$w = \sqrt{\beta^2 - n_2^2 k_0^2} \, a \tag{4.17}$$

where A and B are arbitrary constants, and K_m is the mth-order modified Bessel function of the second kind.

Using the transverse magnetic-field components H_r and H_θ calculated from Eqs. (4.15) and (4.16), and the axial component H_z, we obtain the values of $i_r \times [(1/n_2^2)\nabla \times H]$ on the boundary Γ as follows:

$$i_r \times \left(\frac{1}{n_2^2}\nabla \times H\right)\bigg|_\Gamma = \begin{bmatrix} 0 & 0 \\ \omega\varepsilon_0 K_m(w) & 0 \\ -\dfrac{\omega\varepsilon_0 m\beta a}{w^2} K_m(w) & \dfrac{k_0^2 a}{w} K_m'(w) \end{bmatrix} \begin{bmatrix} \tilde{A} \\ \tilde{B} \end{bmatrix} \tag{4.18}$$

where $\tilde{A} = A\exp(-jm\theta)$, $\tilde{B} = A\exp(-jm\theta)$, and $K_m'(w)$ is the value of a derivative of $K_m(wr/a)$ at $r = a$.

Also, using the values of tangential magnetic-field components H_θ and H_z on the boundary Γ, $\tilde{A}$ and $\tilde{B}$ are expressed as

$$\begin{bmatrix} \tilde{A} \\ \tilde{B} \end{bmatrix} = \begin{bmatrix} -\dfrac{w}{\omega\varepsilon_0 n_2^2 a}\dfrac{1}{K_m'(w)} & \dfrac{m\beta}{\omega\varepsilon_0 n_2^2 w}\dfrac{1}{K_m'(w)} \\ 0 & \dfrac{1}{K_m(w)} \end{bmatrix} \begin{bmatrix} H_\theta(a) \\ H_z(a) \end{bmatrix} . \tag{4.19}$$

Substituting Eq. (4.19) into Eq. (4.18), we obtain

$$i_r \times \left(\frac{1}{n_2^2}\nabla \times H\right)\bigg|_\Gamma = j[Z_0]_\Gamma\{H\}_\Gamma \exp(-jm\theta) \tag{4.20}$$

with

$$[Z_0]_\Gamma = \begin{bmatrix} 0 & 0 & 0 \\ 0 & -\dfrac{w}{n_2^2 a}\dfrac{K_m(w)}{K_m'(w)} & \dfrac{m\beta}{n_2^2 w}\dfrac{K_m(w)}{K_m'(w)} \\ 0 & \dfrac{m\beta}{n_2^2 w}\dfrac{K_m(w)}{K_m'(w)} & -\dfrac{m^2\beta^2 a}{n_2^2 w^3}\dfrac{K_m(w)}{K_m'(w)} + \dfrac{k_0^2 a}{w}\dfrac{K_m'(w)}{K_m(w)} \end{bmatrix} \qquad (4.21)$$

where $\{H\}_\Gamma$ is the nodal magnetic-field vector on the boundary Γ.

4.2.4   Combination of finite element and analytical relations

Substituting Eq. (4.20) into Eq. (4.6), we obtain

$$F = \{H\}^T([K] + s[L] - k_0^2[M])\{H\} + \{H\}_\Gamma^T[Z]_\Gamma\{H\}_\Gamma \qquad (4.22)$$

with

$$[Z]_\Gamma = 2\pi a [Z_0]_\Gamma \,. \qquad (4.23)$$

From the variational principle we obtain the following final matrix equation:

$$[A]\{H\} = \{0\} \qquad (4.24)$$

with

$$\begin{aligned} [A] &= [\tilde{K}] + s[L] - k_0^2[M] \qquad &(4.25) \\ [\tilde{K}] &= [K] + [Z] \qquad &(4.26) \end{aligned}$$

where the matrix $[Z]$ is obtained by adapting the matrix $[Z]_\Gamma$ to the matrix $[K]$.

The condition that Eq. (4.24) has a nontrivial solution is given by

$$|A| = 0 \qquad (4.27)$$

which is the proper equation for the guided modes in an optical fiber.

When setting the artificial boundary in the position far from the core region, we obtain the following eigenvalue problem instead of Eq. (4.24):

$$([K] + s[L])\{H\} - k_0^2[M]\{H\} = \{0\} \,. \qquad (4.28)$$

4.3   Assessment of Vector Finite Element Method

As is well known, an optical fiber will support the propagation of waves having four possible field configurations, classified as the TE_{0n}, TM_{0n}, EH_{mn}, and HE_{mn} modes, where the subscripts m and n are the azimuthal and radial mode numbers, respectively. For a step-index fiber the eigenvalue equation (characteristic equation) can be obtained analytically as

$$\left[\frac{J'_m(u)}{uJ_m(u)} + \frac{K'_m(w)}{wK_m(w)}\right]\left[\left(\frac{n_1}{n_2}\right)^2 \frac{J'_m(u)}{uJ_m(u)} + \frac{K'_m(w)}{wK_m(w)}\right]$$
$$= m^2\left(\frac{1}{u^2} + \frac{1}{w^2}\right)\left[\left(\frac{n_1}{n_2}\right)^2 \frac{1}{u^2} + \frac{1}{w^2}\right] \qquad (4.29)$$

with

$$u = \sqrt{n_1^2 k_0^2 - \beta^2}\, a \qquad (4.30)$$

where n_1 and n_2 are the refractive indices of the core and cladding, respectively, and J_m is the mth-order Bessel function of the first kind.

Now, we consider a graded-core optical fiber with an α-power refractive-index profile given by

$$n(r) = \begin{cases} n_1[1 - 2\Delta(r/a)^\alpha]^{1/2} & 0 \leq r \leq a \\ n_2 = n_1(1 - 2\Delta)^{1/2} & a \leq r \end{cases} \qquad (4.31)$$

with

$$\Delta = \frac{n_1^2 - n_2^2}{2n_1^2} \simeq \frac{n_1 - n_2}{n_1} \qquad (4.32)$$

where Δ is the relative refractive-index difference.

Figure 4.2 shows the propagation characteristics of this fiber calculated from Eq. (4.27) with the penalty coefficient $s = 1/n_2^2$, where n_1=1.515, n_2=1.5, 10 elements and 21 nodes are used, and the normalized frequency v and the normalized propagation constant b are defined as

$$v = n_1 k_0 a\sqrt{2\Delta} = \sqrt{u^2 + w^2} \qquad (4.33)$$

$$b = \frac{(\beta/k_0)^2 - n_2^2}{n_1^2 - n_2^2} \simeq \frac{(\beta/k_0) - n_2}{n_1 - n_2}. \qquad (4.34)$$

The finite element solutions for the step-index ($\alpha = \infty$) fiber shown in Fig. 4.2(a) agree well with the exact solutions calculated from Eq. (4.29) despite near cutoff. The finite element solutions for the parabolic-index (α=2) and triangular-index (α=1) fibers shown, respectively, in Figs. 4.2(b) and (c) also agree well with the high-accuracy solutions calculated from the power-series expansion method (PSEM).[7]

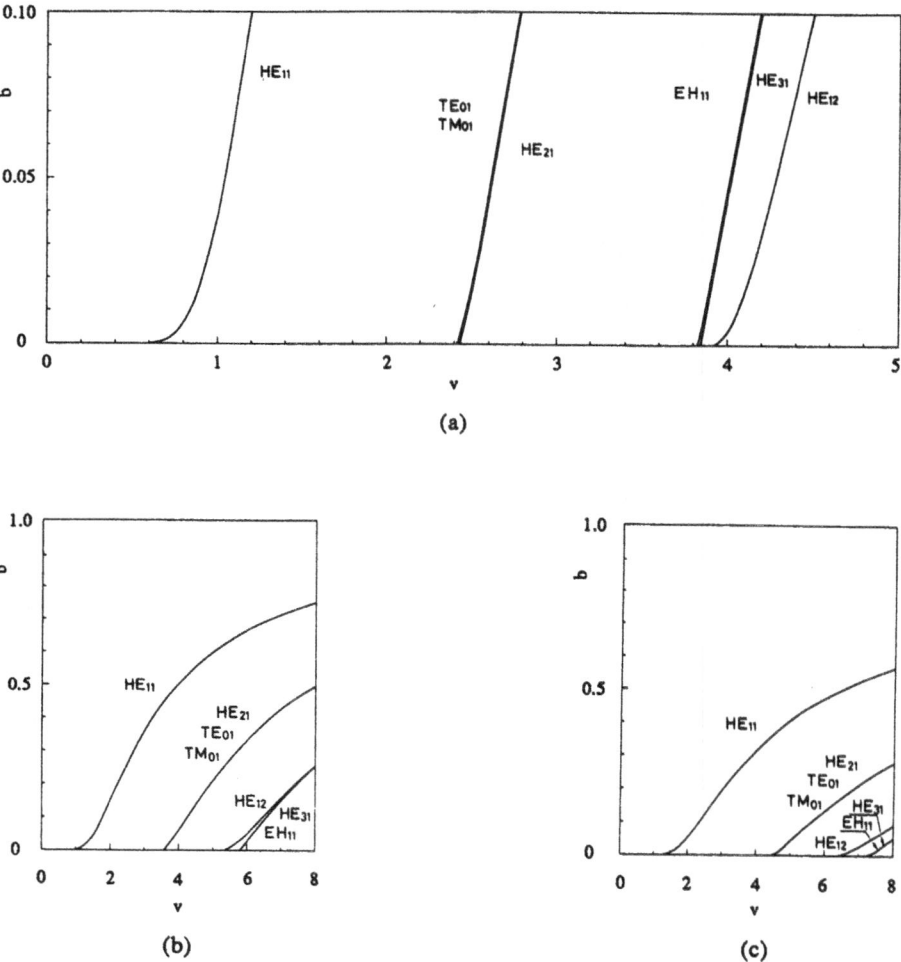

Figure 4.2: Propagation characteristics of an α-power refactive-index optical fiber. (a) $\alpha = \infty$. (b) $\alpha = 2$. (c) $\alpha = 1$.

4.4 Method of Approximate Scalar Wave Analysis

4.4.1 Basic equations

In the weakly guiding approximation ($\Delta \ll 1$), there is a degeneracy of modes, and the TE, TM, EH, and HE modes can be re-expressed in terms of linearly-polarized modes, LP modes, with greatly simplified properties. Table 4.1 illustrates these degeneracies by tabulating the corresponding TE_{0n}, TM_{0n}, EH_{mn}, and HE_{mn} modes for each LP_{ln} mode, where the subscript l is the azimuthal mode number for the LP modes and is related to the azimuthal mode number for the TE, TM, EH, and HE modes as follows:

Table 4.1: Degeneracies of LP modes (including polarization degeneracy).

LP mode	Equivalent mode	Degeneracy
$LP_{0n}(l = 0)$	HE_{1n}	2
$LP_{1n}(l = 1)$	HE_{2n}, TE_{0n}, TM_{0n}	4
$LP_{ln}(l = 2, 3, 4, \cdots)$	$HE_{l+1,n}$, $EH_{l-1,n}$	4

$$l = \begin{cases} 1 & \text{for TE or TM modes} \\ m+1 & \text{for EH modes} \\ m-1 & \text{for HE modes.} \end{cases} \tag{4.35}$$

Propagation characteristics for the LP modes are determined by solving the following scalar wave equation:

$$\frac{1}{r}\frac{d}{dr}\left(r\frac{d\phi}{dr}\right) + \left(k_0^2 n^2 - \beta^2 - \frac{l^2}{r^2}\right)\phi = 0 \tag{4.36}$$

where ϕ is the transverse-field function representing the radial variation of the wave amplitude.

Solving Eq. (4.36), we find the electromagnetic-field components as follows:

(1) TE modes:

$$E_\theta = \phi(r) \tag{4.37a}$$

$$H_r = -(\beta/\omega\mu_0)\phi(r) \tag{4.37b}$$

$$H_z = j(1/\omega\mu_0)(d\phi/dr + \phi/r) \tag{4.37c}$$

$$E_r = E_z = H_\theta = 0. \tag{4.37d}$$

(2) TM modes:

$$E_r = \phi(r) \tag{4.38a}$$
$$E_z = -j(1/\beta)(d\phi/dr + \phi/r) \tag{4.38b}$$
$$H_\theta = (\beta/\omega\mu_0)\phi(r) \tag{4.38c}$$
$$E_\theta = H_r = H_z = 0. \tag{4.38d}$$

(3) EH and HE modes:

$$E_r = \phi(r)\left\{ \begin{array}{c} \cos[(l \mp 1)\theta] \\ \sin[(l \mp 1)\theta] \end{array} \right\} \tag{4.39a}$$

$$E_\theta = \mp\phi(r)\left\{ \begin{array}{c} -\sin[(l \mp 1)\theta] \\ \cos[(l \mp 1)\theta] \end{array} \right\} \tag{4.39b}$$

$$E_z = -j\frac{1}{\beta}\left(\frac{d\phi}{dr} \pm \frac{l}{r}\phi\right)\left\{ \begin{array}{c} \cos[(l \mp 1)\theta] \\ \sin[(l \mp 1)\theta] \end{array} \right\} \tag{4.39c}$$

$$H_r = \pm\frac{\beta}{\omega\mu_0}\phi(r)\left\{ \begin{array}{c} -\sin[(l \mp 1)\theta] \\ \cos[(l \mp 1)\theta] \end{array} \right\} \tag{4.39d}$$

$$H_\theta = \frac{\beta}{\omega\mu_0}\phi(r)\left\{ \begin{array}{c} \cos[(l \mp 1)\theta] \\ \sin[(l \mp 1)\theta] \end{array} \right\} \tag{4.39e}$$

$$H_z = \mp j\frac{1}{\omega\mu_0}\left(\frac{d\phi}{dr} \pm \frac{l}{r}\phi\right)\left\{ \begin{array}{c} -\sin[(l \mp 1)\theta] \\ \cos[(l \mp 1)\theta] \end{array} \right\}. \tag{4.39f}$$

Equation (4.39) is to be interpreted with upper and lower signs corresponding to the EH and HE modes, respectively, whilst the bracketed pairs of trigonometric functions are to be made consistently throughout, but are of course independent of the sign convention for the modes.

Transformation relations for the transverse-field components from polar to Cartesian coordinates are given by

$$E_x = E_r \cos\theta - E_\theta \sin\theta \tag{4.40a}$$
$$E_y = E_r \sin\theta + E_\theta \cos\theta. \tag{4.40b}$$

For a step-index fiber the eigenvalue equation based on Eq. (4.36) can be obtained analytically as

$$\frac{uJ_{l-1}(u)}{J_l(u)} = -\frac{wK_{l-1}(w)}{K_l(w)}. \tag{4.41}$$

Using the normalized frequency v and the normalized propagation constant b, Eq. (4.41) can be rewritten as

$$\frac{J_{l-1}(\sqrt{1-b}\,v)K_l(\sqrt{b}\,v)}{J_l(\sqrt{1-b}\,v)K_{l-1}(\sqrt{b}\,v)} = -\sqrt{\frac{b}{1-b}}. \tag{4.42}$$

Noting $J_{-1}(\sqrt{1-b}\,v) = -J_1(\sqrt{1-b}\,v)$ and $K_{-1}(\sqrt{b}\,v) = K_1(\sqrt{b}\,v)$, the eigenvalue equation for the LP$_{0n}$ modes ($l = 0$) is reduced to

$$\frac{J_1(\sqrt{1-b}\,v)K_0(\sqrt{b}\,v)}{J_0(\sqrt{1-b}\,v)K_1(\sqrt{b}\,v)} = \sqrt{\frac{b}{1-b}} \, . \tag{4.43}$$

The first solution of Eq. (4.43) corresponds to the fundamental LP$_{01}$ mode ($n = 1$) with no cutoff frequency. From Eqs. (4.39) and (4.40) the transverse-field components E_x and E_y for the LP$_{01}$ mode are written as

$$E_x = \phi(r) \tag{4.44a}$$
$$E_y = 0 \tag{4.44b}$$

for the upper function of the bracketed pairs in Eq. (4.39), and

$$E_x = 0 \tag{4.45a}$$
$$E_y = \phi(r) \tag{4.45b}$$

for the lower function of those in Eq. (4.39). Equations (4.44) and (4.45) denote that there is a perfect degeneracy of the two LP$_{01}$ (or HE$_{11}$) modes: both the x-polarized ($E_y = 0$) and y-polarized ($E_x = 0$) LP$_{01}$ modes have the same propagation constant (see Table 4.1). The x-polarized and y-polarized LP$_{01}$ modes are called the HE$_{11}^x$ and HE$_{11}^y$ modes, respectively.

The functional for Eq. (4.36) is given by

$$F = \iint_\Omega \left[\frac{d\phi^*}{dr}\frac{d\phi}{dr} + \left(\frac{l^2}{r^2} + \beta^2 - k_0^2 n^2 \right) \phi^* \phi \right] d\Omega - \int_\Gamma \phi^* \psi \, d\Gamma \tag{4.46}$$

where $\psi = d\phi/dr$.

4.4.2 Finite element approach

Dividing the core region (see Fig. 4.1) into a number of quadratic ring elements (see Fig. 1.5(b)), we expand the transverse-field function ϕ in each element as

$$\phi = \{N\}^{\mathrm{T}}\{\phi\}_e \, . \tag{4.47}$$

Substituting Eq. (4.47) into Eq. (4.46), from the variational principle we obtain the following global matrix equation:

$$[K]\{\phi\} - k_0^2[M]\{\phi\} = \{\psi\} \tag{4.48}$$

with

$$[K] = 2\pi \sum_e \int_e \left[r\{N_r\}\{N_r\}^{\mathrm{T}} + \frac{l^2}{r}\{N\}\{N\}^{\mathrm{T}} \right.$$

$$+ \beta^2 r \{N\}\{N\}^{\mathrm{T}} \bigg] dr \tag{4.49}$$

$$[M] = 2\pi \sum_e \int_e n^2 r \{N\}\{N\}^{\mathrm{T}} dr \tag{4.50}$$

$$\{\psi\} = \begin{bmatrix} 0 \\ 0 \\ \vdots \\ 2\pi a \psi(a) \end{bmatrix} \tag{4.51}$$

where the last component of the $\{\phi\}$ vector is the value of ϕ at node on $r = a$, namely, the core-cladding boundary Γ.

4.4.3   Analytical approach

The transverse-field function ϕ in the cladding region is given by

$$\phi \propto K_l(wr/a) . \tag{4.52}$$

From Eq. (4.52) the value of ψ at $r = a$ is given by

$$\psi(a) = \frac{w K_l'(w)}{a K_l(w)} \phi(a) . \tag{4.53}$$

4.4.4   Combination of finite element and analytical relations

Substituting Eq. (4.53) into Eq. (4.48), we obtain the following final matrix equation:

$$[A]\{\phi\} = \{0\} \tag{4.54}$$

with

$$[A] = [\tilde{K}] - k_0^2 [M] \tag{4.55}$$

$$[\tilde{K}] = [K] + \begin{bmatrix} 0 & 0 & \cdots & & 0 \\ 0 & 0 & \cdots & & 0 \\ \vdots & \vdots & \ddots & & \vdots \\ 0 & 0 & \cdots & & -2\pi \dfrac{w K_l'(w)}{K_l(w)} \end{bmatrix} . \tag{4.56}$$

The condition that Eq. (4.54) has a nontrivial solution is given by

$$|A| = 0 \tag{4.57}$$

which is the proper equation for the LP modes in a weakly guiding optical fiber.

When setting artificial boundaries in the position far from the core region, we obtain the following eigenvalue problem instead of Eq. (4.54):

$$[K]\{\phi\} - \beta^2[M]\{\phi\} = \{0\} \tag{4.58}$$

with

$$[K] = 2\pi \sum_e \int_e \left[n^2 k_0^2 r\{N\}\{N\}^{\mathrm{T}} - r\{N_r\}\{N_r\}^{\mathrm{T}} - \frac{l^2}{r}\{N\}\{N\}^{\mathrm{T}} \right] dr \tag{4.59}$$

$$[M] = 2\pi \sum_e \int_e r\{N\}\{N\}^{\mathrm{T}} dr . \tag{4.60}$$

4.5 Assessment of Scalar Finite Element Method

Single-mode fibers with very low attenuation and sufficient reliability are indispensable for the future ultra-long span optical transmission system. As one promising candidate for such high-grade fiber, a pure-silica-core fiber with fluorine added cladding was developed, and the median attenuation was 0.35 dB/km at 1.3 μm and 0.21 dB/km at 1.55 μm.[8] The minimum attenuation of 0.154 dB/km at 1.55–1.56 μm was achieved.[8]

In addition to silica glass fibers, there is strong interest in fabricating non-silica glass optical fibers which have the potential of achieving much lower losses than silica. Much of the current effort has been focused on zirconium fluoride based glasses which have a predicted minimum loss of 0.01 dB/km in the 2–2.5 μm region.[9] Other ultra low loss (0.001–0.01 dB/km) glasses include heavy metal chlorides, bromides, and iodides with the wavelength of minimum loss generally between 2 and 10 μm.[9] The ultra low loss of these materials implies the possibility of communication systems with very large repeater spacings. In order to utilize these increased repeater spacings, even at relatively low bit rates, the fibers have to be single mode and low dispersion. Using the relative refractive-index difference Δ, the normalized frequency v, and the normalized propagation constant b, the total dispersion (chromatic dispersion) σ is written as

$$\sigma = \sigma_m + \sigma_w \tag{4.61}$$

with

$$\sigma_m = \frac{\lambda}{c} \frac{d^2 n}{d\lambda^2} \tag{4.62}$$

$$\sigma_w = \frac{nv\Delta}{c\lambda} \frac{d^2(vb)}{dv^2} \tag{4.63}$$

where σ_m and σ_w are called the material dispersion and the waveguide dispersion, respectively, and the light velocity c is given by

$$c = 1/\sqrt{\varepsilon_0 \mu_0} \, . \tag{4.64}$$

In order to evaluate the material dispersion, we may make use of refractive index data for fiber materials, which is conventionally expressed in terms of a three-term Sellmeier equation

$$n(\lambda) = \left(1 + \sum_{i=1}^{3} \frac{a_i \lambda^2}{\lambda^2 - \lambda_i^2} \right)^{1/2} \tag{4.65}$$

or a conventional polynomial

$$n(\lambda) = A\lambda^{-4} + B\lambda^{-2} + C + D\lambda^2 + E\lambda^4 \, . \tag{4.66}$$

Table 4.2 shows the refractive-index data coefficients for a pure-silica glass,[10] a zirconium-barium-lanthanum-aluminum fluoride glass (ZBLA),[9] and ZrF_4 - BaF_2 - LaF_3 - YF_3 - AlF_3 - LiF - NaF fluoride glass (ZBLYAL),[11] where λ is in μm. The computed material dispersion spectra are plotted in Fig. 4.3. The ZBLA fluoride glass is typical for most zirconium and hafnium based fluoride glasses, and the minimum attenuation of 0.01 dB/km at 2.2 μm is predicted. It is found from Fig. 4.3 that the material dispersion of the ZBLA glass is zero at 1.684 μm and is -13.9 ps/(km·nm) at the minimum loss wavelength of 2.2 μm.[9]

Table 4.2: Refractive-index data coefficients.

SiO$_2$	ZBLA	ZBLYAL
$a_1 = 0.6961663$	$a_1 = 0.79675$	$A = 2.99073 \times 10^{-5}$
$a_2 = 0.4079426$	$a_2 = 0.49208$	$B = 3.14700 \times 10^{-3}$
$a_3 = 0.8974794$	$a_3 = 0.96404$	$C = 1.50140$
$\lambda_1 = 0.0684043$	$\lambda_1 = 0.085763$	$D = -1.34907 \times 10^{-3}$
$\lambda_2 = 0.1162414$	$\lambda_2 = 0.099847$	$E = -3.74185 \times 10^{-6}$
$\lambda_3 = 9.896161$	$\lambda_3 = 16.00572$	

In this section, using the scalar finite element method (SFEM), we discuss the waveguide design necessary to achieve low dispersion at the minimum loss wavelength. The results of the SFEM will be compared with those of the vector finite element method (VFEM). The FEM is utilized for calculating the normalized propagation constant b as a function of the

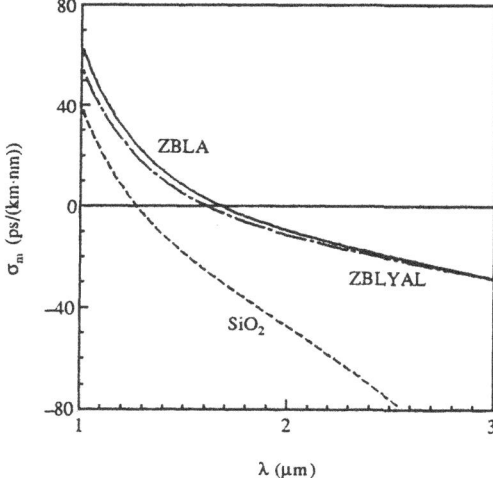

Figure 4.3: Material dispersion spectra.

normalized frequency v (v-b curves), and then the total dispersion is evaluated by repeating numerical differentiation based on the spline interpolation method.

We consider a ZBLA fluoride glass optical fiber with an α-power refractive-index profile (see Eq. (4.31)).

Table 4.3 shows core diameters $2a_{ZTD}$ needed to achieve the zero total dispersion (ZTD) at the minimum loss wavelength of 2.2 μm and the corresponding normalized frequencies v_{ZTD}, where Δ is assumed to be 1%. The dispersion shifted fiber whose ZTD is shifted to the minimum loss wavelength of 2.2 μm behaves as a single-mode fiber at 2.2 μm, because the normalized cutoff frequency v_c of the first higher-order LP_{11} mode is given by[12)]

$$
v_c = \begin{cases} 2.405 & \text{for } \alpha = \infty \\ 3.518 & \text{for } \alpha = 2 \\ 4.381 & \text{for } \alpha = 1 \, . \end{cases} \tag{4.67}
$$

The total dispersion spectra for the fundamental LP_{01} mode are shown in Fig. 4.4. The total dispersion of step-index profile has a smaller slope and is low over a relatively broader wavelength range.

Here, we consider the dispersion shifted fluoride glass fiber operating at 2.2 μm.

Table 4.3: Core diameters $2a_{ZTD}$ needed to achieve the ZTD at 2.2 μm and the corresponding normalized frequencies v_{ZTD}.

α	$2a_{ZTD}$ (μm)		v_{ZTD}	
	SFEM	VFEM	SFEM	VFEM
∞	6.025	6.078	1.846	1.863
2	8.375	8.415	2.567	2.579
1	9.528	9.567	2.920	2.932

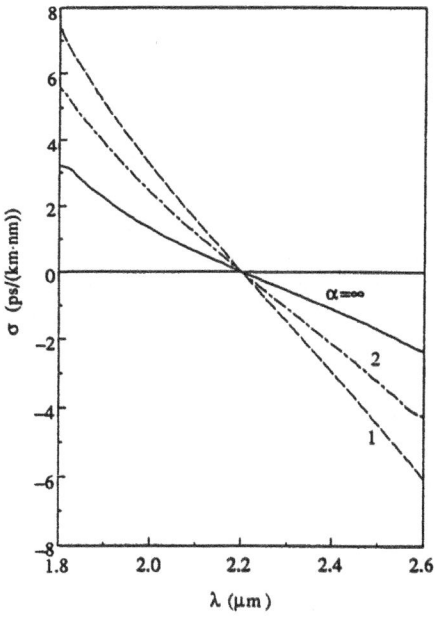

Figure 4.4: Total dispersion spectra.

Table 4.4 shows the mode field diameters (MFD), $2W$, given by[13)]

$$W^2 = \frac{2 \int_0^\infty \phi^2 r\, dr}{\int_0^\infty (d\phi/dr)^2 r\, dr}.$$

(4.68)

Table 4.4: Mode field diameters.

α	MFD (μm)	
	SFEM	VFEM
∞	8.016	8.030
2	8.861	8.868
1	10.099	10.104

Figure 4.5 shows the offset losses α_d given by[14)]

$$\alpha_d = -10 \log \eta$$

(4.69)

with

$$\eta = \frac{\left(\iint_S \phi_1 \phi_2\, dS \right)^2}{\iint_S \phi_1^2\, dS \iint_S \phi_2^2\, dS}$$

(4.70)

where d is the offset value, $\iint_S dS$ denotes the integration over a cross section of butt-jointed fibers with an offset misalignment, and the subscripts 1 and 2 designate the quantities in input and output fibers, respectively.

Table 4.5 shows the allowable bending radii R_b, at which the bending loss exceeds a given value, where, considering the minimum loss of the ZBLA glass, namely, 0.01 dB/km, the permissible bending loss is assumed to be 0.001 dB/km. The bending loss α_b is given by[15)]

$$\alpha_b = \frac{\sqrt{\pi}}{4} \frac{a[\phi(a)/K_0(w)]^2 \exp\left(-\frac{4w^3 \Delta}{3av^2}R\right)}{w \int_0^\infty \phi^2 r\, dr \left(\frac{wR}{a} + \frac{v^2}{2w\Delta}\right)^{1/2}}$$

(4.71)

where R is the radius of curvature.

The modal field ϕ for the LP$_{01}$ mode calculated from the SFEM is utilized to evaluate Eqs. (4.68) to (4.71). In the VFEM the transverse magnetic-field distribution in the radial direction for the HE$_{11}$ mode, namely, $H_r(r)$ or $H_\theta(r)$, is used in place of the transverse-field function ϕ in Eqs. (4.68) to (4.71).

It is found from Tables 4.3 to 4.5 that the results of the SFEM (20 elements and 41 nodes are used) agree well with those of the VFEM (20 elements and 41 nodes are used).

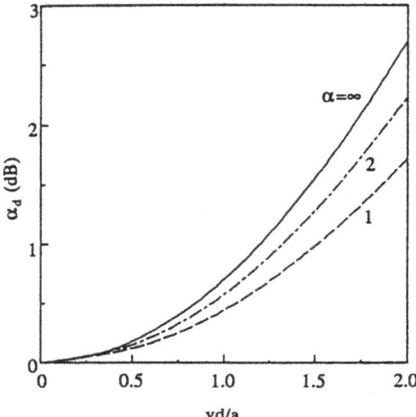

Figure 4.5: Offset losses.

Table 4.5: Allowable bending radii.

α	R_b (cm)	
	SFEM	VFEM
∞	1.257	1.253
2	1.857	1.850
1	3.379	3.365

References

1) K. Okamoto and T. Okoshi, "Vectorial wave analysis of inhomogeneous optical fibers using finite element method", *IEEE Trans. Microwave Theory Tech.*, Vol MTT-26, No. 2, pp. 109–114, Feb. 1978.

2) C. C. Su, "Origin of spurious modes in the analysis of optical fiber using the finite-element or finite-difference technique", *Electron. Lett.*, Vol. 21, No. 21, pp 858–860, Sept. 1985.

3) K. Okamoto, "Comparison of calculated and measured impulse responses of optical fibers", *Appl. Opt.*, Vol. 18, No. 13, pp. 2199–2206, July 1979.

4) T. A. Lenahan, "Calculation of modes in an optical fiber using the finite element method and EISPACK", *Bell Syst. Tech. Jour.*, Vol. 62, No. 9, pp. 2663–2694 Nov. 1983.

5) C. C. Su, "Eigenproblems of radially inhomogeneous optical fibers from the scalar formulation", *IEEE Jour. Quantum Electron.*, Vol. QE-21, No. 10, pp. 1554–1557, Oct. 1985.

6) Q. Y. Li, "Propagation characteristics of single-mode optical fibers with arbitrary refractive-index profile: The finite quadratic element approach", *IEEE/OSA Jour Lightwave Technol.*, Vol. 9, No. 1, pp. 22–26, Jan. 1991.

7) K. Oyamada and T. Okoshi, "High-accuracy numerical data of dispersion and group-delay characteristics of α-power graded-core fibers", *IEEE Trans. Microwave Theory Tech.*, Vol. MTT-28, No. 10, pp. 1113–1118, Oct. 1980.

8) H. Kanamori, H. Yokota, G. Tanaka, M. Watanabe, Y. Ishiguro, I. Yoshida T. Kakii, S. Itoh, Y. Asano, and S. Tanaka, "Transmission characteristics and reliability of pure-silica-core single-mode fibers", *IEEE/OSA Jour. Lightwave Technol.*, Vol. LT-4, No. 8, pp. 1144–1150, Aug. 1986.

9) K. L. Walker, M. M. Broer, and A. Carnevale, "Dispersion shifted fiber designs with low bending losses for infrared materials", *SPIE Proc.*, Vol. 618, pp. 17–24 1986.

10) I. H. Mallitson, "Interspecimen comparison of the refractive index of fused silica" *Jour. Opt. Soc. Am.*, Vol. 55, No. 10, pp. 1205–1209, Oct. 1965.

11) Y. Ohishi and S. Takahashi, "Low-dispersion fluoride glass single-mode fibres operating in two spectral ranges", *Electron. Lett.*, Vol. 24, No. 4, pp. 220–221, Feb. 1988.

12) W. A. Gambling, D. N. Payne, and H. Matsumura, "Cut-off frequency in radially inhomogeneous single-mode fibre", *Electron. Lett.*, Vol. 13, No. 5, pp. 139–140 March 1977.

13) K. Petermann, "Constraints for fundamental-mode spot size for broadband dispersion-compensated single-mode fibres", *Electron. Lett.*, Vol. 19, No. 18 pp. 712–714, Sept. 1983.

14) J. Sakai and T. Kimura, "Splice loss evaluation for optical fibers with arbitrary-index profile", *Appl. Opt.*, Vol. 17, No. 17, pp. 2848–2853, Sept. 1978.

15) J. Sakai and T. Kimura, "Bending loss of propagation modes in arbitrary-index profile optical fibers", *Appl. Opt.*, Vol. 17, No. 10, pp. 1499–1506, May 1978.

Chapter 5

POLARIZATION-MAINTAINING OPTICAL FIBERS

5.1 Introduction

Actual fibers generally exhibit some ellipticity of a core and/or some anisotropy in the refractive index distribution due to anisotropic stresses. This results in two different propagation constants for the x-polarized and y-polarized LP_{01} modes, namely, HE_{11}^x and HE_{11}^y modes (see Subsection 4.4.1), leading to perturbations of the state of polarization (SOP) of the light transmitted by the fiber.

Polarization-maintaining optical fibers, which can maintain the SOP and can transmit the phase information efficiently, are of great interest for use in fiber-optic sensing systems and coherent optical communication systems.[1],[2] Linear polarization-maintaining fibers are classified into two groups. One is the axially-nonsymmetrical type making use of geometrical birefringence. The other is the stress-applied type making use of stress-induced birefringence. Typical examples of the former are side-pit and side-tunnel fibers, and the latter ones are bow-tie and PANDA (Polarization-maintaining AND Absorption-reducing) fibers.[1],[2] In order to estimate the transmission characteristics of such fibers, electromagnetic-field analyses are required.[3]−[14]

In this chapter, to evaluate the characteristics of the polarization-maintaining optical fiber, the FEM is used both for the modal analysis[3],[5]−[10],[12]−[14] and for the stress analysis,[4],[9]−[11],[14] which can easily be applied to the fiber having an arbitrary refractive-index profile. For the modal analysis, both vector and scalar finite element formulations are presented.

5.2   Method of Modal Analysis

5.2.1   Vector finite element method

In the case of axially nonsymmetrical fibers making use of geometrical birefringence such as side-pit and side-tunnel fibers, their refractive-index distributions in the cross section become complex, and further weakly-guiding approximations are no longer applicable due to the abrupt refractive-index variation at the interface between the core and the tunnels. Therefore, their rigorous analysis becomes extremely difficult.

Among many approaches, a vector H-field FEM with the penalty variational principle (see Subsection 3.3.2), namely, Eq. (3.30), enables one to analyze the optical waveguidance in such complex geometries and refractive-index profiles. Equation (3.30) is the generalized eigenvalue problem whose eigenvalue is k_0^2 and eigenvector is $\{H\}$, where k_0 is the wavenumber of free space and $\{H\}$ is the nodal magnetic-field vector. In Eq. (3.30), spurious solutions do not appear in the region $\beta/k_0 \geq 1/\sqrt{s}$ (see Eq. (3.29)), where β is the phase constant in the propagation direction and s is the penalty coefficient.

In the FEM analysis of the open-type waveguide like an optical fiber, the problem is how to treat an infinite region. Although several kinds of methods have been proposed, the most popular one is that introducing artificial boundaries, which is widely used. However, in order to get accurate results even near cutoff and for higher-order modes, it is necessary to put this artificial boundary far enough from a core.

Here, for a step-index circular-core optical fiber, taking one case in which the artificial boundaries are introduced and another case in which the infinite elements (see Subsection 1.6.3) are added outside the region divided into ordinary finite elements, we will compare and examine the accuracy of solutions in each case.

Making use of the symmetry nature of the system, we subdivide only one-quarter of the fiber cross section into linear triangular elements (see Fig. 1.3(a)) as shown in Fig. 5.1, where N_E, N_P, and N_I are the numbers of elements, nodes, and infinite elements, respectively. The boundary conditions on the planes of symmetry are

$$H_x \;=\; 0 \qquad\qquad \text{on } AB \qquad\qquad (5.1a)$$

$$H_x \;=\; H_z = 0 \qquad \text{on } BC \qquad\qquad (5.1b)$$

for the HE_{11} mode, and

$$H_x \;=\; 0 \qquad \text{on } AB \qquad\qquad (5.2a)$$

$$H_y \;=\; 0 \qquad \text{on } BC \qquad\qquad (5.2b)$$

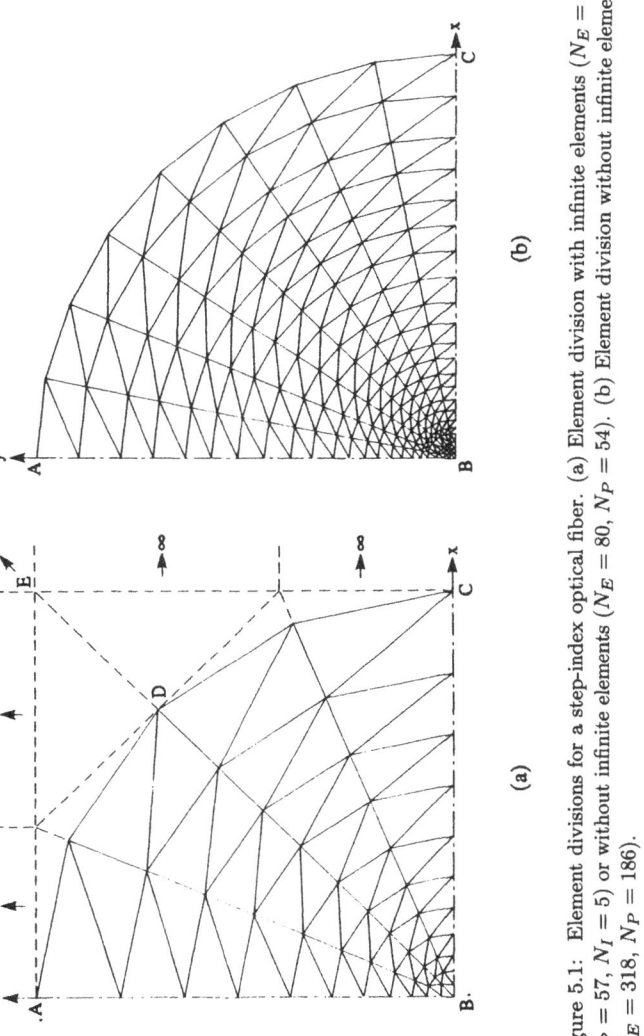

Figure 5.1: Element divisions for a step-index optical fiber. (a) Element division with infinite elements ($N_E = 86$, $N_P = 57$, $N_I = 5$) or without infinite elements ($N_E = 80$, $N_P = 54$). (b) Element division without infinite elements ($N_E = 318$, $N_P = 186$).

for the TE_{01} mode. A zero boundary, $H_x = H_y = H_z = 0$, is assumed on AC since $a \ll R$, where a is the core radius, R is the distance between a core center and an artificial-zero boundary, and $a/R = 0.03$.

Table 5.1 shows the results calculated from the element division of Fig. 5.1(a), where the refractive index of cladding $n_2 = 1.46$, the relative refractive-index difference $\Delta = 1\%$, the penalty coefficient $s = 1$, v is the normalized frequency, b is the normalized propagation constant, and the relative error e is defined by

$$e = (v_{\text{FEM}} - v_{\text{exact}})/v_{\text{exact}} . \tag{5.3}$$

Here v_{FEM} and v_{exact} are the finite element and exact solutions, respectively.

Table 5.1: Influence of an artificial-zero boundary and effect of infinite elements on the solutions.

	HE$_{11}$ mode ($\beta a = 6.0$)		TE$_{01}$ mode ($\beta a = 19.0$)	
	v	e	v	e
With infinite element	0.857076	4.4×10^{-3}	2.714127	8.1×10^{-2}
Without infinite element	0.857085	5.5×10^{-3}	2.714127	8.1×10^{-2}
Exact solution	0.857038		2.711935	
Corresponding b-value	0.011976		0.084941	

It is found from Table 5.1 that, in the case of $a/R = 0.03$, despite near cutoff (i.e., b is nearly zero), the results for the case using finite elements only, that is, the case setting an artificial-zero boundary, and those for the case adding infinite elements are almost the same both for the fundamental (HE_{11}) and the higher-order (TE_{01}) modes. Therefore, in all the following analyses, we will put the artificial-zero boundary on the position $a/R = 0.03$, where we use the boundary condition $\boldsymbol{H} = 0$.

First, to check the validity of the results by the vector FEM, we will consider a step-index circular-core optical fiber whose exact solutions can be obtained easily and examine the accuracy of calculations.

Figure 5.2 shows the dispersion characteristics for the HE_{11} and TE_{01} modes of a step-index circular-core optical fiber. It is found from Fig. 5.2 that we can raise the accuracy of solutions monotonically and systematically by increasing the number of elements.

Also shown in Fig. 5.2 are the results that have been obtained using the extrapolation method to improve the accuracy of solutions in terms of two original results. Putting the solutions for $N_E = 80$ and $N_E = 318$ to $k_{0,1}^2$

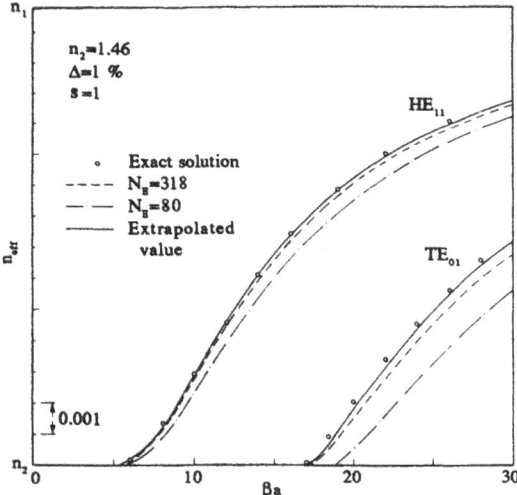

Figure 5.2: Vector finite element solution of a step-index circular-core optical fiber.

and $k_{0,2}^2$, respectively, the extrapolated value $k_{0,\text{ext}}^2$ is derived:

$$k_{0,\text{ext}}^2 = \frac{k_{0,2}^2 - h^2 k_{0,1}^2}{1 - h^2}, \quad h = \sqrt{\frac{80}{318}} \approx \frac{1}{2}. \tag{5.4}$$

It is found from Fig. 5.2 that greater accuracy is possible by using Eq. (5.4), which may be useful for the reduction of computer memory and CPU time.

Next, let us comment on the penalty coefficient s included in Eq. (3.30), because the accuracy of solutions depends on the value of s. Generally, the larger the value of s becomes, the more stable the solutions become, but the worse their accuracy. Also, on the contrary, the smaller the value of s becomes, the better the accuracy of physical solutions tends to become, but then the spurious solutions occur in the guided region $\beta/k_0 \geq n_2$, and those spurious solutions may couple to physical solutions, with the result that the accuracy of the physical solutions may be rather worse. Here, we will describe a method that keeps the solutions stable and makes them more accurate.

Now, consider s_1 and s_2 ($s_1 \neq s_2$) as s. Letting the solutions (eigenvalues) for $s = s_1$ and $s = s_2$ to be $k_0^2(s_1)$ and $k_0^2(s_2)$, respectively, and extrapolating linearly, the solution $k_0^2(0)$ at $s = 0$ is estimated:

$$k_0^2(0) = \frac{s_1 k_0^2(s_2) - s_2 k_0^2(s_1)}{s_1 - s_2}. \tag{5.5}$$

Selecting suitable values for s_1 and s_2, we can obtain accurate results by using Eq. (5.5) without suffering from the spurious solutions. According to Eq. (3.29), the spurious solutions do not occur in the guided region $\beta/k_0 \geq n_2$ by setting $s \geq 1/n_2^2$, so we set $s_1 = 1/n_2^2$ and $s_2 = 1$ here.

Figure 5.3 shows the relative errors for the fundamental and the first higher-order modes taking s as a parameter. (The definition of the relative error e is the same as that in Eq. (5.3).) The smaller the value of s becomes, the smaller the error becomes monotonically, which verifies the usefulness of Eq. (5.5).

5.2.2 Scalar finite element method

In ordinary single-mode fibers, the scalar wave approximation (LP approximation) is applicable, since the relative refractive-index difference is sufficiently small. However, the earlier approximate scalar FEM (see Section 3.5) was developed for waveguides for use in optical integrated circuits and/or diode lasers and therefore cannot be applied directly to the analysis of optical fibers. Accordingly, we shall reformulate an alternative version of the approximate scalar FEM for the analysis of weakly guiding anisotropic fibers such as bow-tie and PANDA fibers. Although two approaches are possible, in which either the electric field or the magnetic field is taken as a leading function, the former approach is adopted here.

Consider an anisotropic medium having a diagonal permittivity tensor, and approximate the HE_{11}^x and HE_{11}^y modes by the modes $E_y \equiv 0$ and $E_x \equiv 0$, respectively, since E_x and H_y are dominant in the former, whereas E_y and H_x are dominant in the latter. From Maxwell's equations the following Helmholtz equation is derived:

$$p_x \frac{\partial^2 \phi}{\partial x^2} + p_y \frac{\partial^2 \phi}{\partial y^2} - \beta^2 \phi + q k_0^2 \phi = 0 \tag{5.6}$$

where the field ϕ and the coefficients p_x, p_y, q are written as follows:

(1) HE_{11}^x modes:

$$\phi = E_x \tag{5.7}$$
$$p_x = n_x^2/n_z^2 \tag{5.8a}$$
$$p_y = 1 \tag{5.8b}$$
$$q = n_x^2 . \tag{5.8c}$$

(2) HE_{11}^y modes:

$$\phi = E_y \tag{5.9}$$
$$p_x = 1 \tag{5.10a}$$
$$p_y = n_y^2/n_z^2 \tag{5.10b}$$
$$q = n_y^2 . \tag{5.10c}$$

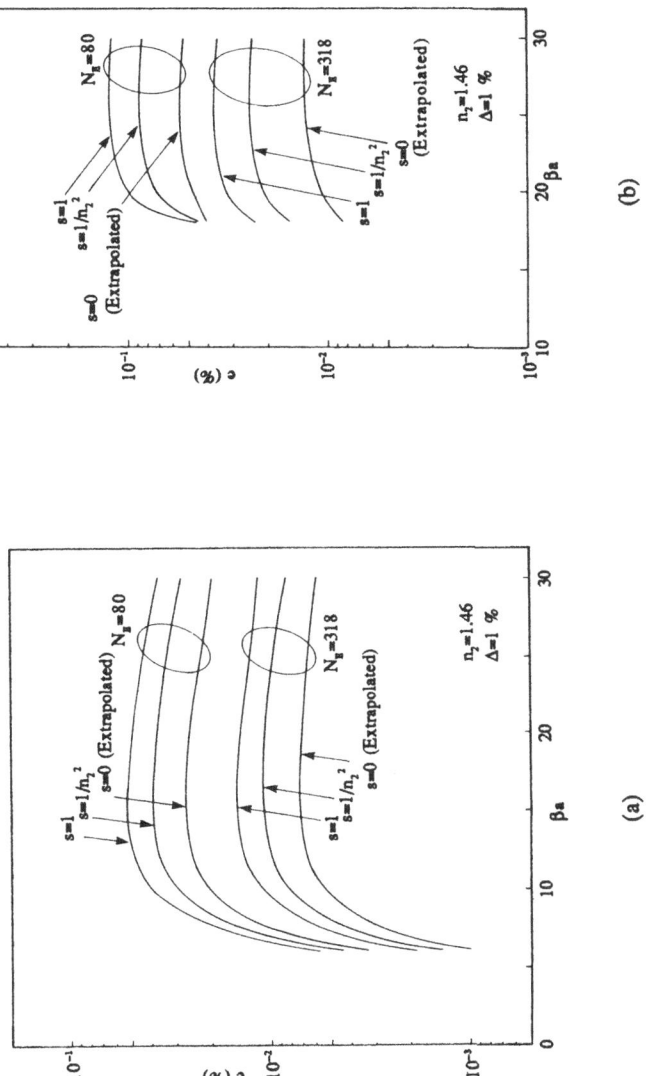

Figure 5.3: Relative errors of solutions. (a) HE_{11} mode. (b) TE_{01} mode.

Dividing the fiber cross section into a number of linear triangular elements (see Fig. 1.3(a)) and applying the standard finite element procedure, we obtain the following eigenvalue problem:

$$[K]\{\phi\} - \beta^2[M]\{\phi\} = \{0\} \tag{5.11}$$

with

$$[K] = \sum_e \iint_e [qk_0^2\{N\}\{N\}^{\mathrm{T}} - p_x\{N_x\}\{N_x\}^{\mathrm{T}}$$

$$- p_y\{N_y\}\{N_y\}^{\mathrm{T}}]\, dx\, dy \tag{5.12}$$

$$[M] = \sum_e \iint_e \{N\}\{N\}^{\mathrm{T}}\, dx\, dy. \tag{5.13}$$

To verify the validity of the formulation, let us apply Eq. (5.11) to a circular step-index fiber and compare the results obtained with exact solutions. Figure 5.4 shows the dispersion characteristics for the HE_{11} mode of a step-index circular-core optical fiber. As is evident from Fig. 5.4, the present results are in good agreement with exact calculations.

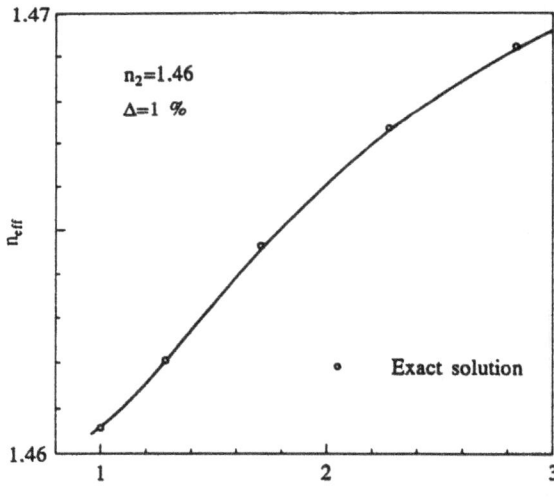

Figure 5.4: Scalar finite element solution of a step-index circular-core optical fiber.

5.3   Method of Stress Analysis

Stress-applied optical fibers have birefringence (anisotropy) through the photoelastic effect. The refractive indices for the x, y, and z directions, denoted by n_x, n_y, and n_z, respectively, are given by

$$
\begin{bmatrix} n_x \\ n_y \\ n_z \end{bmatrix} = \begin{bmatrix} n_{x0} \\ n_{y0} \\ n_{z0} \end{bmatrix} - \begin{bmatrix} C_1 & C_2 & C_2 \\ C_2 & C_1 & C_2 \\ C_2 & C_2 & C_1 \end{bmatrix} \begin{bmatrix} \sigma_x \\ \sigma_y \\ \sigma_z \end{bmatrix} \tag{5.14}
$$

where σ_x, σ_y, and σ_z are the normal stress components for the x, y, and z directions, respectively, C_1 and C_2 are the photoelastic coefficients, and n_{x0}, n_{y0}, and n_{z0} are the refractive indices of the unstressed fiber.

Here, we present a stress analysis by the FEM[4] to enable us to find the refractive-index distribution of a fiber cross section.

As in the case of optical fibers, when the dimension of the structure in one direction (z direction) is very large in comparison with the other two directions (x and y directions) and the applied forces act in the xy plane and do not vary in the z direction, the problem is called the plane strain problem.

The functional for the plane strain problem is expressed by using the strain energy per unit length in the direction of propagation as

$$
F = \iint_\Omega \boldsymbol{S}^* \cdot \boldsymbol{T} \, dx \, dy \tag{5.15}
$$

with

$$
\boldsymbol{S} = \begin{bmatrix} S_{xx} \\ S_{yy} \\ 2S_{xy} \end{bmatrix} \tag{5.16}
$$

$$
\boldsymbol{T} = \begin{bmatrix} T_{xx} = \sigma_x \\ T_{yy} = \sigma_y \\ T_{xy} \end{bmatrix} \tag{5.17}
$$

where $\iint_\Omega dx \, dy$ denotes the integration over the fiber cross section Ω, S_{xx}, S_{yy} are the normal strains, and S_{xy} and T_{xy} are the shear strain and stress, respectively.

The strain-displacement relation is given by

$$
\boldsymbol{S} = \nabla_s \boldsymbol{u} \tag{5.18}
$$

with

$$
\nabla_s = \begin{bmatrix} \partial/\partial x & 0 \\ 0 & \partial/\partial y \\ \partial/\partial y & \partial/\partial x \end{bmatrix} \tag{5.19}
$$

$$
\boldsymbol{u} = \begin{bmatrix} u_x \\ u_y \end{bmatrix} \tag{5.20}
$$

where u_x and u_y are the x and y components of the particle displacement $\boldsymbol{u}$, respectively.

For the plane strain problem the stress-strain relation is expressed as

$$\boldsymbol{T} = [c](\boldsymbol{S} - \boldsymbol{S}_0) \tag{5.21}$$

with

$$[c] = \begin{bmatrix} \dfrac{(1-\nu)E}{(1+\nu)(1-2\nu)} & \dfrac{\nu E}{(1+\nu)(1-2\nu)} & 0 \\ \dfrac{\nu E}{(1+\nu)(1-2\nu)} & \dfrac{(1-\nu)E}{(1+\nu)(1-2\nu)} & 0 \\ 0 & 0 & \dfrac{E}{2(1+\nu)} \end{bmatrix} \tag{5.22}$$

$$\boldsymbol{S}_0 = \begin{bmatrix} (1+\nu)\alpha\Delta T \\ (1+\nu)\alpha\Delta T \\ 0 \end{bmatrix} \tag{5.23}$$

where E is Young's modulus, ν is Poisson's ratio, α is the thermal-expansion coefficient, ΔT is the temperature change (negative on cooling), and $\boldsymbol{S}_0$ is called the initial strain.

Substituting Eqs. (5.18) and (5.21) into Eq. (5.15), we obtain

$$F = \iint_\Omega (\nabla_s \boldsymbol{u})^* \cdot [c](\nabla_s \boldsymbol{u} - \boldsymbol{S}_0)\, d\Omega . \tag{5.24}$$

Dividing the fiber cross section into a number of linear triangular elements (see Fig. 1.3(a)), we expand the particle displacement $\boldsymbol{u}$ in each element as

$$\boldsymbol{u} = [N]^{\mathrm{T}}\{u\}_e \tag{5.25}$$

with

$$\{u\}_e = \begin{bmatrix} \{u_x\}_e \\ \{u_y\}_e \end{bmatrix} \tag{5.26}$$

$$[N] = \begin{bmatrix} \{N\} & \{0\} \\ \{0\} & \{N\} \end{bmatrix} \tag{5.27}$$

where $\{u_x\}_e$ and $\{u_y\}_e$ are the nodal particle displacement vectors.

Substituting Eq. (5.25) into Eq. (5.24) and assuming the constant initial strain in each element, from the variational principle we obtain the following simultaneous linear equation (deterministic problem):

$$[K]\{u\} = \{f_T\} \tag{5.28}$$

with

$$[K] = \sum_e \iint_e [B][c][B]^{\mathrm{T}} \, dx \, dy \qquad (5.29)$$

$$\{f_T\} = \sum_e \iint_e [B][c]\boldsymbol{S}_0 \, dx \, dy \qquad (5.30)$$

$$[B] = \begin{bmatrix} \{N_x\} & \{0\} & \{N_y\} \\ \{0\} & \{N_y\} & \{N_x\} \end{bmatrix} \qquad (5.31)$$

where $\{f_T\}$ is the global initial force vector.

Equation (5.28) gives the particle displacements at all nodes of the fiber under thermal stress. Once the global particle displacement vector $\{u\}$ is determined, the stress in each element is given by

$$\boldsymbol{T} = [c]([B]^{\mathrm{T}}\{u\}_e - \boldsymbol{S}_0) \,. \qquad (5.32)$$

5.4 Side-Tunnel Polarization-Maintaining Fibers

5.4.1 Fiber structure

In side-tunnel polarization-maintaining fibers shown in Fig. 5.5(a) relatively high birefringence can be obtained without applying stress through the hollow tunnels at both sides of the core, where n_1, n_2, and n_3 are the refractive indices of the core, cladding, and tunnels, respectively. Furthermore, it is also possible, with these fibers, to have absolute single polarization transmission.

5.4.2 Modal analysis

Making use of the symmetry nature of the system, we subdivide one-quarter of the fiber cross section into isoparametric quadratic elements (see Subsection 1.6.1) as shown in Fig. 5.5(b).

Figure 5.6 shows the modal birefringence at the cutoff v-value of the HE_{11}^x mode, where $n_1 = 1.515$, $n_2 = 1.5$, $n_3 = 1.0$, $r_1/r_2 = 0.15$, v is the normalized frequency, and the modal birefringence B is defined as

$$B = (\beta_y - \beta_x)/k_0 \,. \qquad (5.33)$$

Here β_x and β_y are the phase constants of the HE_{11}^x and HE_{11}^y modes, respectively. The solid line denotes the result obtained by Eq. (5.5) ($s_1 = 1/n_2^2$, $s_2 = 1.0$), and the dotted line denotes the conventional result ($s_1 = 1/n_2^2$). It is found from Fig. 5.6 that the result obtained by Eq. (5.5) is in good agreement with the reliable result by the point-matching method (PMM).[15]

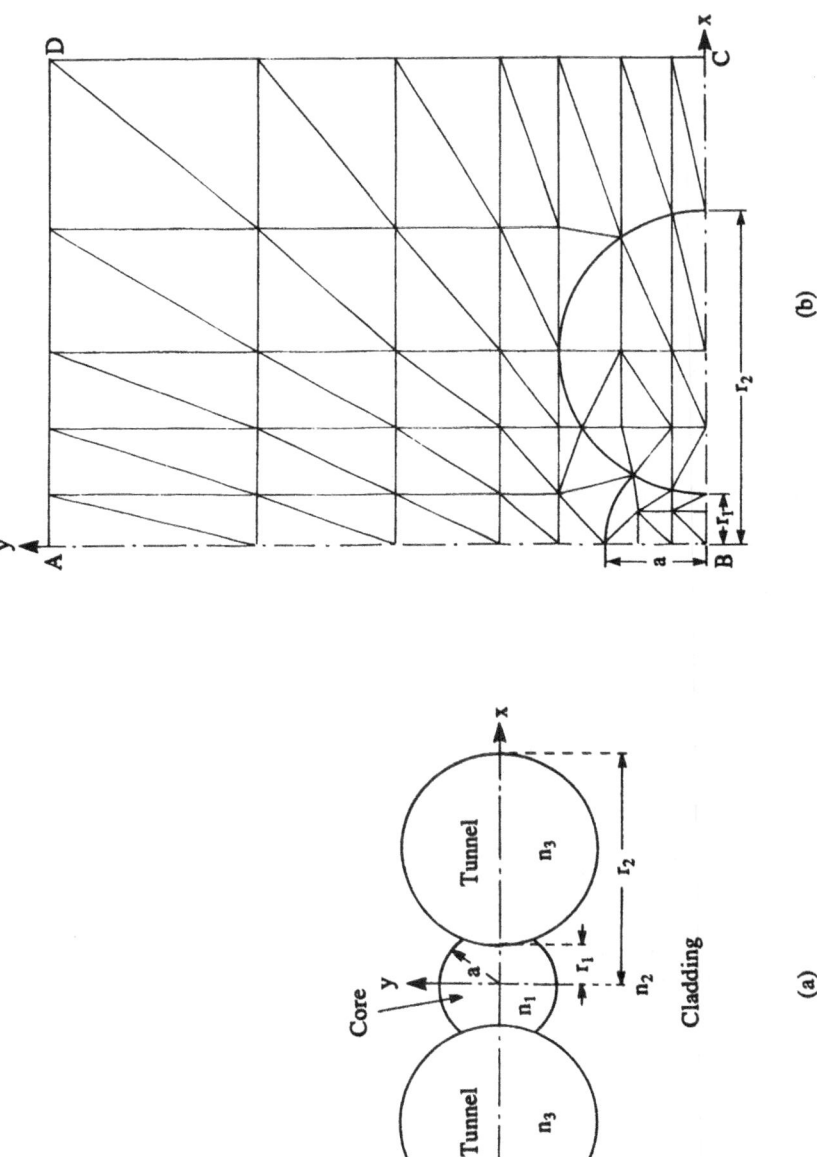

(b)

(a)

Figure 5.5: Side-tunnel polarization-maintaining optical fiber. (a) Fiber structure. (b) Element division.

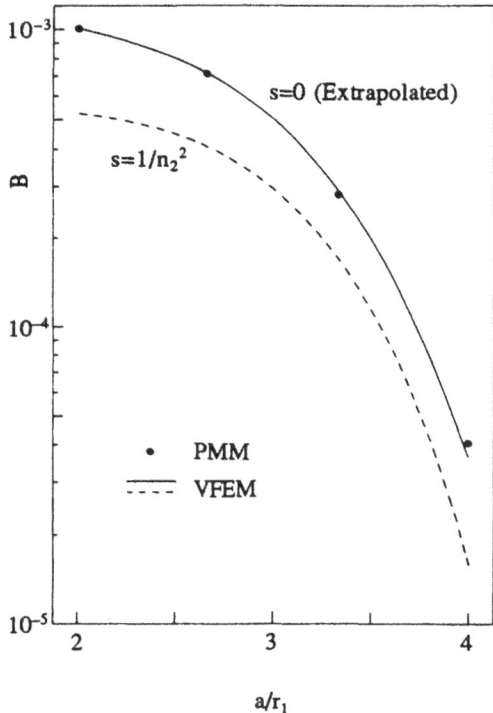

Figure 5.6: Modal birefringence at the cutoff v-value of the HE_{11}^x mode.

Figures 5.7 to 5.9 show the propagation characteristics, the modal bire-fringence, and the polarization-mode dispersion for the HE_{11}^x mode of side-tunnel fibers having various cross-sectional shape, where the polarization-mode dispersion σ_p is defined as

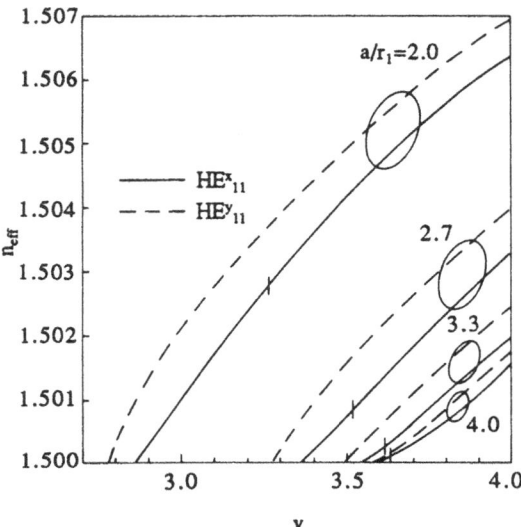

Figure 5.7: Propagation characteristics of a side-tunnel fiber.

$$\sigma_p = d(k_0 B)/dk_0 . \tag{5.34}$$

In these figures, we take $n_1 = 1.515$, $n_2 = 1.5$, and $n_3 = 1.0$ as the values of refractive indices, and change the fiber geometry by varying a/r_1 such as $a/r_1 = 2.0$, 2.7, 3.3, and 4.0. As for the ratio r_1/r_2, we fix on $r_1/r_2 = 0.15$ since the magnitude of r_2 does not greatly affect the modal propagation characteristics compared with that of r_1. The following points are concluded from these results:

1. The cutoff v-values for the fundamental and the higher-order modes are both fairly high in comparison with those for conventional weakly-guiding fibers (| denotes the cutoff for the first higher-order mode).

2. As a/r_1 becomes larger, i.e., the tunnels approach the core center, the propagation constant for the higher-order mode comes to approach that for the fundamental modes.

3. The smaller a/r_1 is, the higher the modal birefringence is. That is, in order to stabilize the polarization state, one should push the tunnels away from the core center and make their size larger.

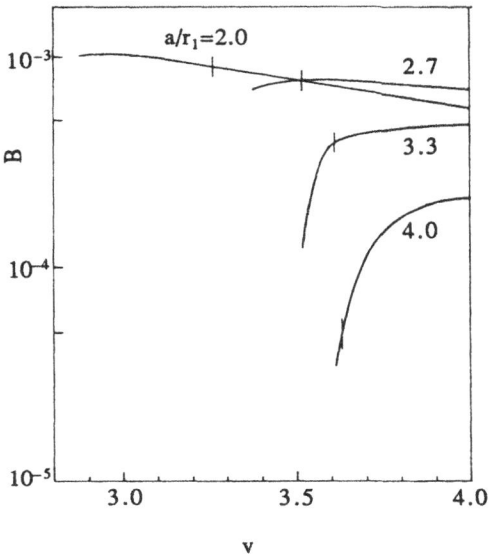

Figure 5.8: Modal birefringence of a side-tunnel fiber.

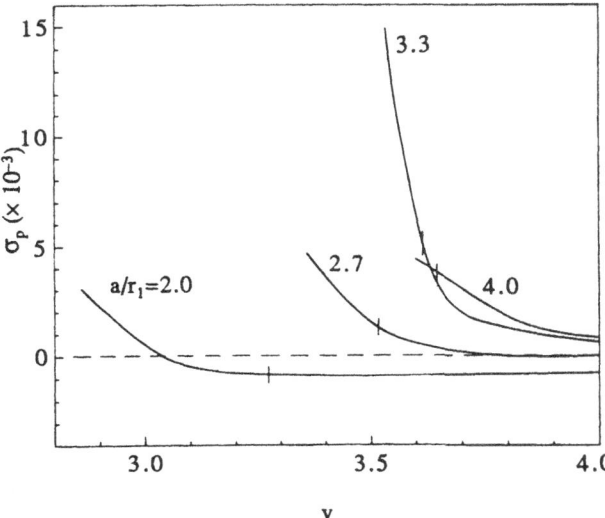

Figure 5.9: Polarization-mode dispersion of a side-tunnel fiber.

4. The magnitude of polarization-mode dispersion σ_p increases abruptly with increases in a/r_1, while zero polarization-mode dispersion ($\sigma_p = 0$) can be realized when $a/r_1 = 2.0$.

As was indicated in the discussion on Fig. 5.8, in the case of side-tunnel fibers, the phenomenon that the propagation constant of the higher-order mode approaches that of the fundamental modes is observed for some cross-sectional shape. We will try to explain this strange phenomenon in the following.

Figures 5.10 and 5.11 display principal magnetic-field components for the fundamental and first higher-order modes, respectively, where n_{eff} is the effective index and the ordinate in each case is normalized by the corresponding maximum value. Although two fundamental modes exist in a fiber which polarize mainly along the x and y directions, respectively, we consider the former, namely, the HE_{11}^x mode. As is evident from the value of effective index n_{eff}, all the fields displayed in Figs. 5.10 and 5.11 are evaluated near cutoff. It is seen from Figs. 5.10 and 5.11 that the fields are sharply varied near the interface between the core and the tunnels, so that the spots have considerably higher ellipticity compared with those of stress-applied fibers.[8] Also, it is very interesting to note that, as the tunnels encroach on the core region (i.e., a/r_1 becomes larger), the field distribution of the fundamental mode is fairly deformed and breaks up into a double peak due to the influence of the hollow tunnels. This results in a close resemblance in shape between the fundamental and the first higher-order modes, as is clearly seen if we compare Figs. 5.11(a) and (b). Hence, both phase constants become very similar as a/r_1 becomes larger.

5.4.3   Stress analysis

For the convenience of element division, assume that the shape of the tunnels is a sector, as shown in Fig. 5.12(a), whose central angle is 90°. The fiber parameters are listed in Table 5.2. Figure 5.12(b) shows the element division for the system of Fig. 5.12(a) and Fig. 5.13 shows the results of analyses. Here B_0 is the stress-induced birefringence B_s at the core center and is given by

$$B_0 = (C_2 - C_1)[\sigma_y(0,0) - \sigma_x(0,0)] . \qquad (5.35)$$

$B_{s,ave}$ and $B_{g,max}$ are the average of B_s in the core and the maximum of geometrical birefringence B_g in the single-mode region, respectively. The following points are concluded from these results:

1. The residual stress in side-tunnel fibers makes their birefringence increase since B_s has the same sign as B_g, i.e., $B_s > 0$ when $B_g > 0$.

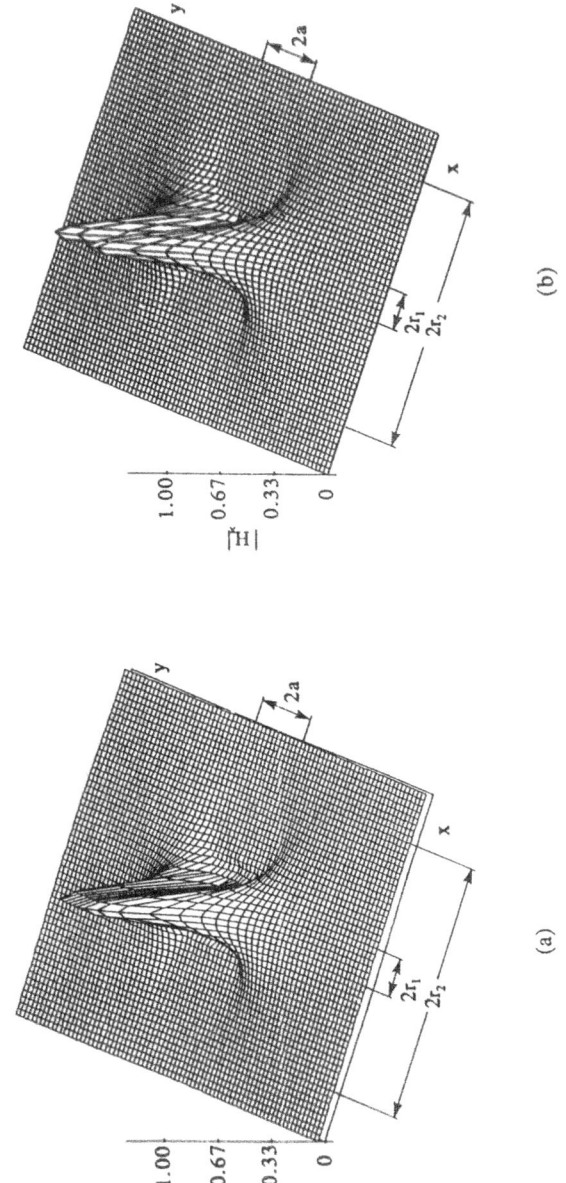

Figure 5.10: Field distributions of principal magnetic-field components in a side-tunnel fiber with $a/r_1 = 2.0$ ($r_1/r_2 = 0.15$). (a) The fundamental mode ($v = 2.88$, $n_{\text{eff}} = 1.50001$). (b) The first higher-order mode ($v = 3.34$, $n_{\text{eff}} = 1.50009$).

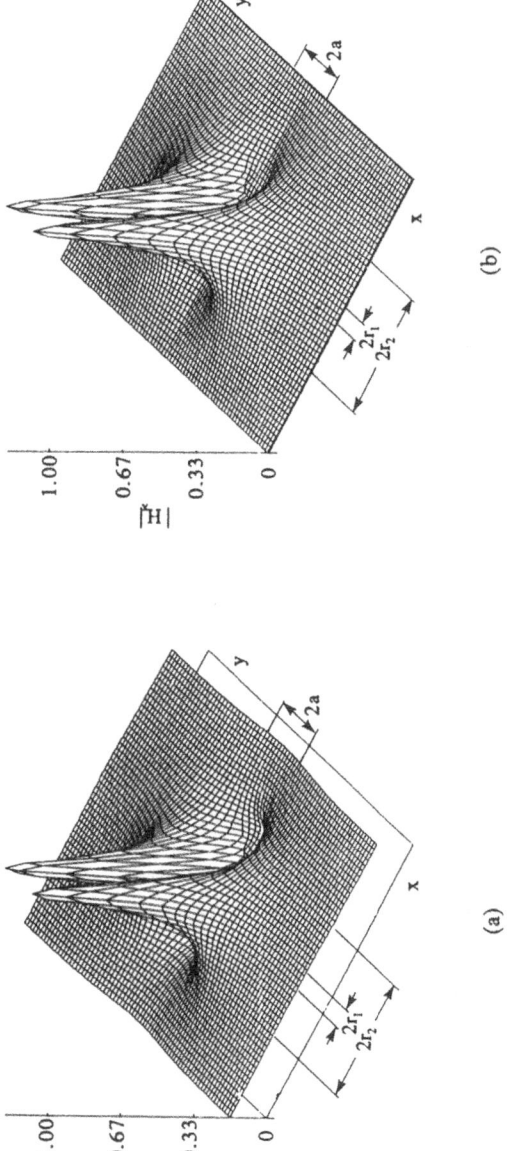

Figure 5.11:   Field distributions of principal magnetic-field components in a side-tunnel fiber with $a/r_1 = 4.0$ ($r_1/r_2 = 0.15$). (a) The fundamental mode ($v = 3.62$, $n_{\text{eff}} = 1.50002$). (b) The first higher-order mode ($v = 3.71$, $n_{\text{eff}} = 1.50006$).

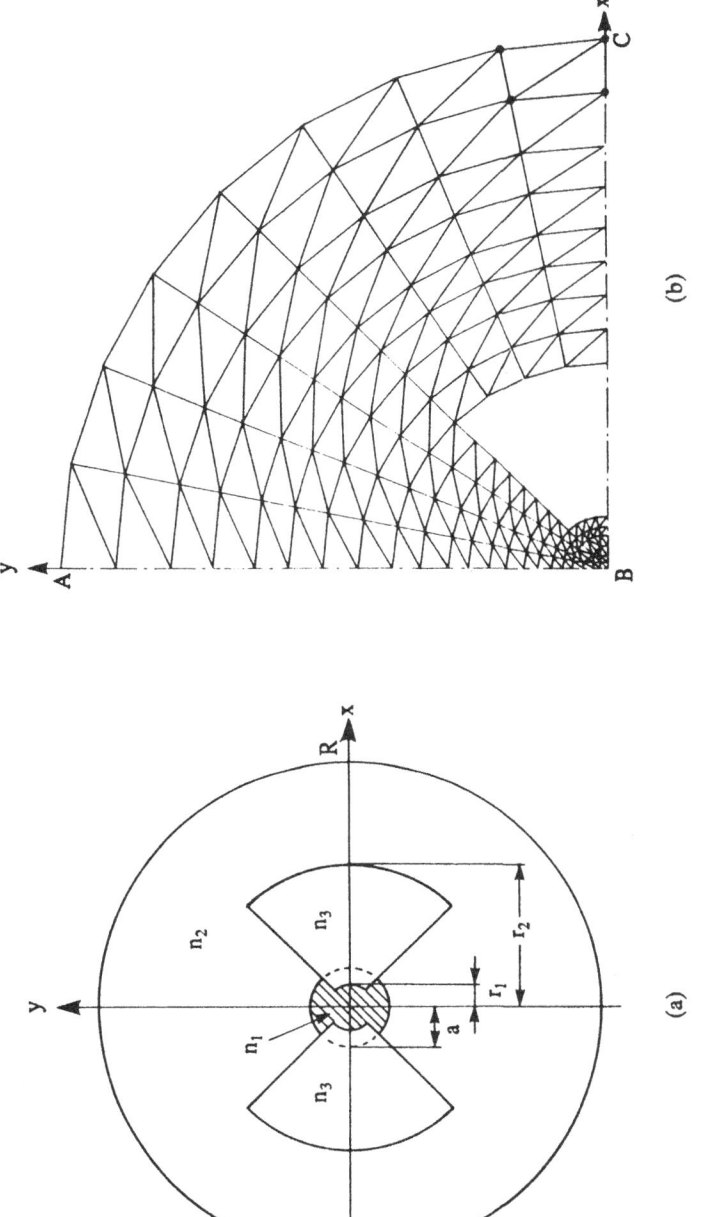

Figure 5.12: Polarization–maintaining fiber with sector tunnels. (a) Fiber structure. (b) Element division.

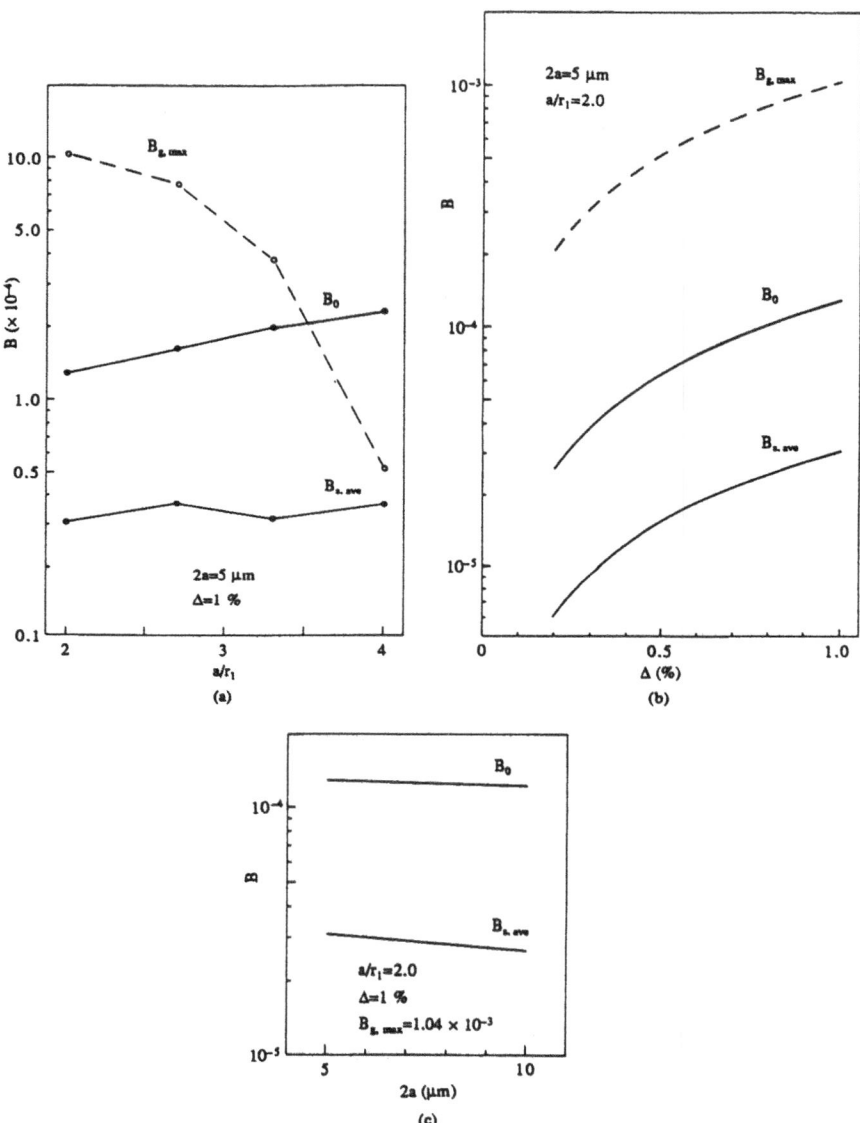

Figure 5.13: Birefringence of a side-tunnel fiber. (a) a/r_1 dependence. (b) Δ dependence. (c) $2a$ dependence.

Table 5.2: Fiber parameters (side-tunnel fiber).

Core diameter	$2a = 5 \sim 10 \ \mu$m
Cladding diameter	$2R = 125 \ \mu$m
Aspect ratio of tunnels	$r_1/r_2 = 0.15$, $a/r_1 = 2.0 \sim 4.0$
Relative refractive-index difference	$\Delta = 0.2 \sim 1.0\%$
Young's modulus	$E = 7830 \ \text{kg/mm}^2$
Poisson's ratio	$\nu = 0.186$
Thermal-expansion coefficient	$\alpha_1 = \alpha_2 + 7.6 \times 10^{-7}\Delta(\%) \ \text{deg}^{-1}$ in core
	$\alpha_2 = 5.4 \times 10^{-7} \ \text{deg}^{-1}$ in cladding
Temperature change	$\Delta T = -1000$ deg
Photoelastic coefficient	$C_1 = 7.421 \times 10^{-6} \ \text{mm}^2/\text{kg}$
	$C_2 = 4.104 \times 10^{-5} \ \text{mm}^2/\text{kg}$

2. B_g decreases rapidly as a/r_1 increases. On the contrary, B_s is not sensitive to a/r_1.

3. B_s is dominant for large a/r_1 while B_g is dominant for small a/r_1.

4. B_s hardly changes with the core diameter.

5. B_s increases monotonically with Δ, and then $B_s \propto \Delta$.

6. The magnitude of B_0 is fairly different from that of $B_{s,\text{ave}}$, which is not the case of the stress-applied type, where $B_0 \simeq B_{s,\text{ave}}$.[9]

5.5 Stress-Applied Polarization-Maintaining Fibers

5.5.1 Fiber structure

Figure 5.14 shows the cross section of a graded-core stress-applied polarization-maintaining fiber, where a and R are the radii of the core and cladding, respectively, and r_1 and r_2 are the inner and outer radii of the stress-applying parts (SAP), respectively. It is assumed that the shape of the SAPs is a sector whose central angle is $90°$. It is also assumed that the refractive-index profile is an α power and that the bulk refractive indices of the core center, cladding, and SAPs are n_1, n_2, and n_3, respectively. Relative refractive-index differences between the core center and the cladding and between the SAPs and the cladding are defined by $\Delta = (n_1^2 - n_2^2)/(2n_1^2)$ and $\Delta_s = (n_3^2 - n_2^2)/(2n_3^2)$, respectively. Fiber parameters are listed in Table 5.3. When α is infinite, the fiber shown in Fig. 5.14 is reduced to the step-core stress-applied fiber.

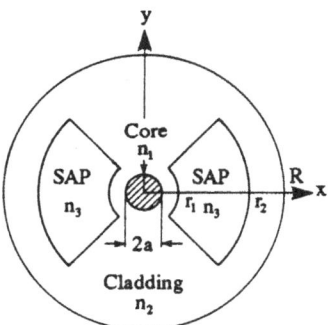

Figure 5.14: Graded-core stress-applied polarization-maintaining optical fiber.

Table 5.3: Fiber parameters (stress-applied fiber).

Core diameter	$2a = 5\ \mu\mathrm{m}$
Cladding diameter	$2R = 125\ \mu\mathrm{m}$
Inner radius of SAPs	$r_1/a = 1.0, 3.0, 5.4$
Outer radius of SAPs	$r_2/R = 0.76$
Refractive-index profile	$n(r) = n_1[1 - 2\Delta(r/a)^\alpha]^{1/2}$ in core
	$= n_2 = 1.458$ in cladding
Relative refractive-index difference	$\Delta = 0.5\%,\ \Delta_s = -0.5\%$
Young's modulus	$E = 7830\ \mathrm{kg/mm^2}$
Poisson's ratio	$\nu = 0.186$
Thermal-expansion coefficient	$\alpha(r) = \alpha_1[1 - 2\Delta'(r/a)^\alpha]^{1/2}\mathrm{deg}^{-1}$ in core
	$= \alpha_2 = 5.4 \times 10^{-7}\mathrm{deg}^{-1}$ in cladding
	$= \alpha_3 = 1.554 \times 10^{-6}\ \mathrm{deg}^{-1}$ in SAPs
	$\alpha_1 = \alpha_2 + 11.04 \times 10^{-7}\Delta(\%)\ \mathrm{deg}^{-1}$
	$\Delta' = (\alpha_1^2 - \alpha_2^2)/(2\alpha_1^2)$
Temperature change	$\Delta T = -800\ \mathrm{deg}$
Photoelastic coefficient	$C_1 = 7.421 \times 10^{-6}\ \mathrm{mm^2/kg}$
	$C_2 = 4.104 \times 10^{-5}\ \mathrm{mm^2/kg}$

5.5.2   Stress analysis

First, to find the refractive-index variation induced by the residual stress, we implement a stress analysis by the FEM. Because of the twofold symmetry of the system, one-quarter of the fiber cross section is subdivided into a number of linear triangular elements (see Fig. 1.3(a)) as shown in Fig. 5.1(b). As is shown in Table 5.3, it is assumed that thermal-expansion coefficient in the graded core has the same profile as the refractive index.

Figure 5.15 shows the refractive-index profile along the x-axis for step-core ($\alpha = \infty$), parabolic-core ($\alpha = 2$), and triangular-core ($\alpha = 1$) profiles. It is found that in and near the core the relation $n_x > n_y$ holds and birefringence is induced in the fiber.

Figure 5.16 displays an example of a refractive-index tomogram for a triangular-core profile ($\alpha = 1$). It is found from Fig. 5.16(b) that strong birefringence peaks are observed around the core and outside the SAPs and that the sign of birefringence is inverted in the SAPs. The result of Fig. 5.16(b) will be important for the determination of the modal cutoff below.

Figure 5.17 shows the variation of birefringence at the core center versus the normalized SAP distance for two extreme cases, i.e., $\alpha = 1$ (triangular) and $\alpha = \infty$ (step). It is evident from Fig. 5.17 that as the SAPs approach the core, the birefringence exponentially increases, and that the characteristics scarcely depend on the refractive-index profile in the core. The following relation has been obtained by the optimum curve fitting:

$$B_0 = 4.42 \times 10^{-4} \exp[-0.3(r_1/a - 1.0)^{0.8}]. \qquad (5.36)$$

5.5.3   Modal analysis

To examine the dependence of modal birefringence on the normalized frequency, we perform modal analysis by the approximate scalar FEM.

Figure 5.18 shows the dispersion characteristics for the two fundamental modes of the graded-core stress-applied fibers taking α as a parameter. Owing to the birefringence induced in the fiber, the refractive index at the surface of the cladding is different and depends on the polarized direction. The lower bounds which determine cutoff are therefore not coincident between the x- and y-polarized modes.

Figure 5.19 exhibits the modal birefringence B against the normalized frequency v for $\alpha =1$, 2, and ∞ (step) taking the normalized SAP distance r_1/a as a parameter. It is found from Fig. 5.19 that, when the SAPs approach the core, the modal birefringence B increases remarkably. In Fig. 5.19 the broken lines denote the region in which the y-polarized mode becomes leaky, while the x-polarized mode remains purely guided. It is interesting to note that, compared with the conventional step-index

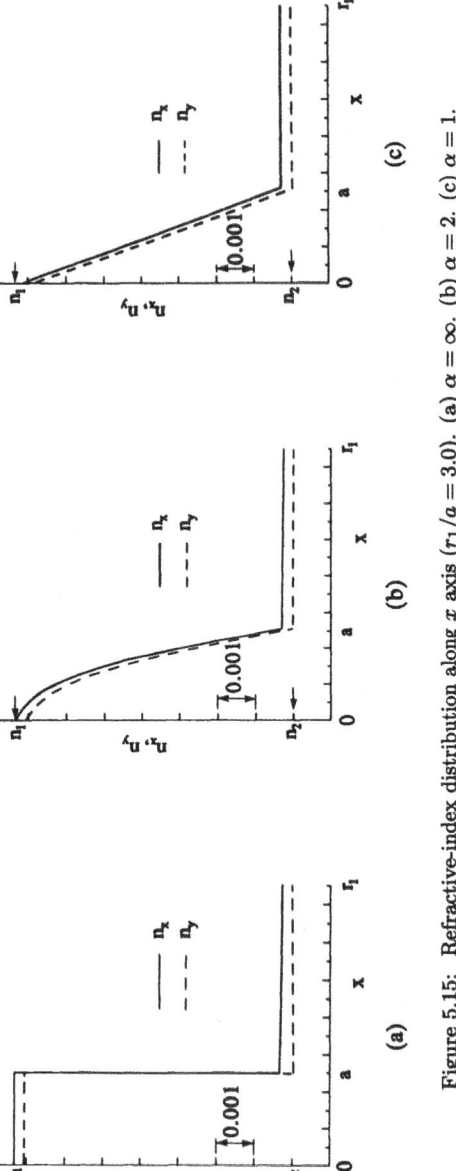

Figure 5.15: Refractive-index distribution along x axis ($r_1/a = 3.0$). (a) $\alpha = \infty$. (b) $\alpha = 2$. (c) $\alpha = 1$.

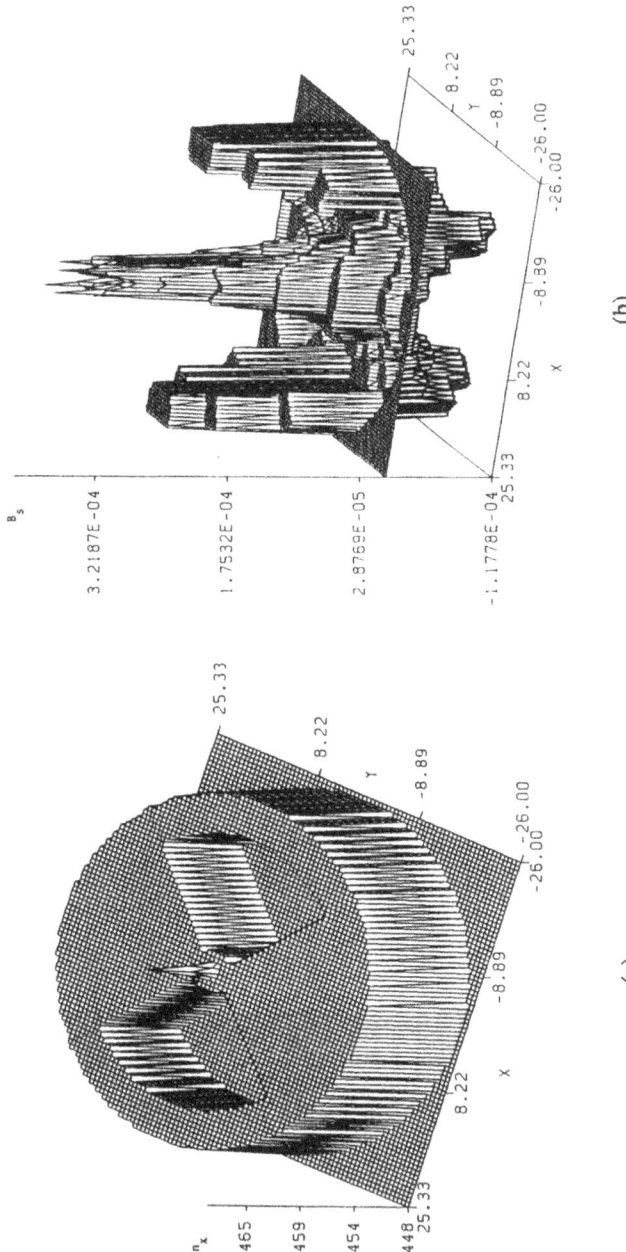

(b)

(a)

Figure 5.16: Refractive-index tomogram for a triangular-core profile $(r_1/a = 3.0)$. (a) n_x. (b) $B_s = n_x - n_y$.

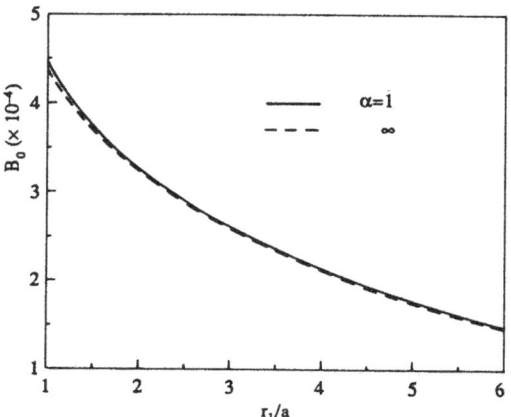

Figure 5.17: Stress-induced birefringence in a core center.

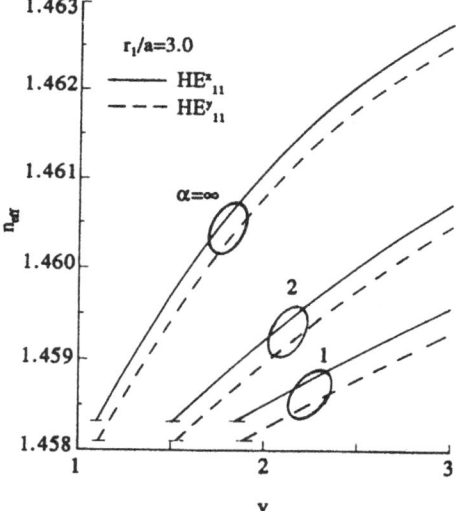

Figure 5.18: Propagation characteristics for a graded-core stress-applied fiber.

profile,[16] i.e., $\alpha = \infty$, the absolutely single-mode bandwidth (the region corresponding to the broken lines) can be considerably enlarged in the case of the graded-core profile, especially for the triangular profile.

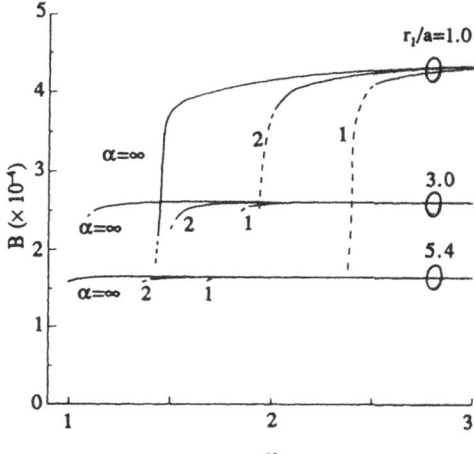

Figure 5.19: Modal birefringence of a graded-core stress-applied fiber.

References

1) T. Okoshi and K. Kikuchi, *Coherent Optical Fiber Communications*, KTK Scientific Publishers (KTK), Tokyo, 1988.

2) T. Okoshi, "Single-polarisation single-mode optical fibers", *IEEE Jour. Quantum Electron.*, Vol. QE-17, No. 6, pp. 879–884, June 1981.

3) T. Okoshi and K. Oyamada, "Single-polarisation single-mode optical fibre with refractive-index pits on both sides of core", *Electron. Lett.*, Vol. 16, No. 16, pp. 712–713, Aug. 1980.

4) K. Okamoto, T. Hosaka, and T. Edahiro, "Stress analysis of optical fibers by a finite element method", *IEEE Jour. Quantum Electron.*, Vol. QE-17, No. 10, pp. 2123–2129, Oct. 1981.

5) K. Oyamada and T. Okoshi, "Two-dimensional finite-element method calculation of propagation characteristics of axially nonsymmetrical optical fibers", *Radio Sci.*, Vol. 17, No. 1, pp. 109–116, Jan.–Feb. 1982.

6) T. Okoshi, K. Oyamada, M. Nishimura, and H. Yokota, "Side-tunnel fiber: An approach to polarization-maintaining optical waveguiding scheme", *Electron. Lett.*, Vol. 18, No. 19, pp. 824–826, Sept. 1982.

7) T. Okoshi, T. Aihara, and K. Kikuchi, "Prediction on the ultimate performance of side-tunnel single-polarization fibre", *Electron. Lett.*, Vol. 19, No. 25/26, pp. 1080–1082, Dec. 1983.

8) K. Hayata, M. Koshiba, and M. Suzuki, "Modal spot size of axially nonsymmetrical fibres", *Electron. Lett.*, Vol. 22, No. 3, pp. 127–129, Jan. 1986.

9) K. Hayata, M. Koshiba, and M. Suzuki, "Vectorial wave analysis of stress-applied polarization-maintaining optical fibers by the finite-element method", *IEEE/OSA Jour. Lightwave Technol.*, Vol. LT-4, No. 2, pp. 133–139, Feb. 1986.

10) K. Hayata, M. Koshiba, and M. Suzuki, "Polarization characteristics of graded-core stress-applied polarization-maintaining fibres", *Electron. Lett.*, Vol. 22, No. 7, pp. 363–365, March 1986.

11) K. Hayata, M. Koshiba, and M. Suzuki, "Stress-induced birefringence of side-tunnel type polarization-maintaining fibers", *IEEE/OSA Jour. Lightwave Technol.*, Vol. LT-4, No. 6, pp. 601–607, June 1986.

12) K. Hayata, M. Eguchi, M. Koshiba, and M. Suzuki, "Anomaly of modal field in side-tunnel single-polarization fibre", *Electron. Lett.*, Vol. 22, No. 16, pp. 838–839, July 1986.

13) K. Hayata, M. Eguchi, M. Koshiba, and M. Suzuki, "Vectorial wave analysis of side-tunnel type polarization-maintaining optical fibers by variational finite-elements", *IEEE/OSA Jour. Lightwave Technol.*, Vol. LT-4, No. 8, pp. 1090–1096, Aug. 1986.

14) K. Hayata and M. Koshiba, "Characteristics of graded-core stress-applied polarization-maintaining single-mode fibers", *Jour. Opt. Soc. Am. A*, Vol. 5, No. 4, pp. 535–541, April 1988.

15) T. Hosono, T. Hinata, and H. Yoshikawa, "Polarization-maintaining optical fibers with hollow circular pits", *IEEE/OSA Jour. Lightwave Technol.*, Vol. LT-4, No. 11, pp. 1609–1616, Nov. 1986.

16) T. Hosaka, K. Okamoto, T. Miya, Y. Sasaki, and T. Edahiro, "Low-loss single polarization fibres with asymmetrical strain birefringence", *Electron. Lett.*, Vol. 17, No. 15, pp. 530–531, July 1981.

Chapter 6

OPTICAL GRATINGS

6.1 Introduction

Diffraction of a plane wave from a periodic structure[1] is of considerable interest in several diverse areas, such as acousto-optics, integrated optics, and spectroscopy. In particular, groove-type (surface-relief or corrugated) dielectric gratings are of great interest owing to their many applications, such as distributed-feedback lasers, waveguide couplers, and spectral filters. Several methods, therefore, have been proposed for the analysis of diffraction characteristics of dielectric gratings.

The above-mentioned dielectric gratings are the transmitted-type gratings. The metallic gratings, on the other hand, are the reflected-type gratings and are also of great interest in reflectors, beam shapers, multiplexers, and other applications. Various methods for the analysis of metallic gratings have been proposed. Some of these methods treat a metallic grating as a perfectly conducting grating. However, in the optical region, the influence of metallic loss on the reflection characteristics cannot be neglected. Recently, some attempts have been made to analyze lossy gratings. These methods were originally developed for the analysis of lossless dielectric gratings and have been extended to lossy gratings. However, it is in general difficult to handle an arbitrarily shaped grating, and thus the FEM has been utilized for the analysis of dielectric and metallic gratings with arbitrary profiles.[2]−[7]

In this chapter a numerical approach based on the FEM is described for the analysis of plane-wave diffraction from groove-type dielectric and metallic gratings. Both cases of the TE and TM mode incidences are systematically formulated. Further, for metallic gratings, a simple method in which an approximate boundary condition using the surface impedance is combined with the FEM is also introduced.

6.2   Method of Analysis

6.2.1   Basic equations

We consider a dielectric grating with period d in the x direction as shown in Fig. 6.1, where n_1 and n_2 are the refractive indices of the input side (incident or reflected side) and of the output side (transmitted side), respectively, the boundaries Γ_1 and Γ_2 are located at $y = y_1$ and $y = y_2$, respectively, and the boundaries Γ_3 and Γ_4 are located at $x = 0$ and $x = d$, respectively.

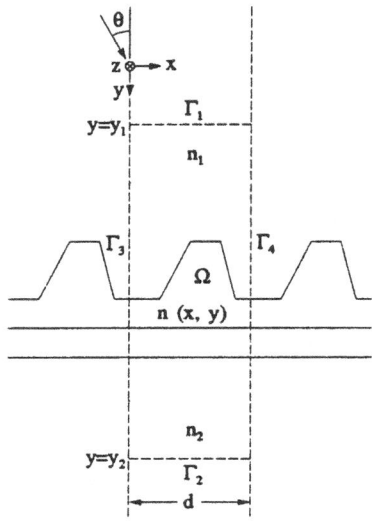

Figure 6.1: Dielectric grating.

Assuming that there is no variation in the z direction, from Maxwell's equation we obtain the following Helmholtz equation:

$$\nabla_t \cdot (p\nabla_t \phi) + q k_0^2 \phi = 0 \qquad (6.1)$$

with

$$\nabla_t = i_x \partial/\partial x + i_y \partial/\partial y \qquad (6.2)$$

where i_x and i_y are the unit vectors in the x and y directions, respectively, and the field ϕ and the coefficients p, q are written as follows:

(1) TE modes:

$$\phi \;=\; E_z \qquad (6.3)$$

$$p \;=\; 1 \qquad (6.4a)$$

$$q \;=\; n^2 \,. \qquad (6.4b)$$

(2) TM modes:

$$\phi = H_z \tag{6.5}$$
$$p = 1/n^2 \tag{6.6a}$$
$$q = 1. \tag{6.6b}$$

We define ψ on Γ_1 to Γ_4 as follows:

$$\psi_1 = -\partial\phi/\partial y \quad \text{on } \Gamma_1 \tag{6.7a}$$
$$\psi_2 = \partial\phi/\partial y \quad \text{on } \Gamma_2 \tag{6.7b}$$
$$\psi_3 = -\partial\phi/\partial x \quad \text{on } \Gamma_3 \tag{6.7c}$$
$$\psi_4 = \partial\phi/\partial x \quad \text{on } \Gamma_4. \tag{6.7d}$$

It is assumed that the TE or TM mode is incident from the negative y direction at an angle θ from the y axis. Then, from Floquet's theorem the periodic boundary conditions are given by

$$\phi_4 = \sigma\phi_3 \tag{6.8}$$
$$p_4\psi_4 = -\sigma p_3\psi_3 \tag{6.9}$$

with

$$\sigma = \exp(-jn_1 k_0 d \sin\theta). \tag{6.10}$$

The functional for Eq. (6.1) is given by

$$
F = \iint_\Omega [p(\nabla_t\phi)^* \cdot (\nabla_t\phi) - qk_0^2\phi^*\phi]\, dx\, dy
$$
$$
- \int_{\Gamma_1} p_1\phi_1^*\psi_1\, d\Gamma - \int_{\Gamma_2} p_2\phi_2^*\psi_2\, d\Gamma
$$
$$
- \int_{\Gamma_3} p_3\phi_3^*\psi_3\, d\Gamma - \int_{\Gamma_4} p_4\phi_4^*\psi_4\, d\Gamma. \tag{6.11}
$$

6.2.2 Finite element approach

Dividing the region Ω into a number of quadratic triangular elements (see Fig. 1.3(b)), we expand the field ϕ in each element as

$$\phi = \{N\}^{\mathrm{T}}\{\phi\}_e. \tag{6.12}$$

Substituting Eq. (6.12) into Eq. (6.11), from the variational principle we obtain the following global matrix equation:

$$
[A]\{\phi\} = \sum_e{}' \int_e \{N\}_1\, p_1\psi_1\, dx + \sum_e{}' \int_e \{N\}_2\, p_2\psi_2\, dx
$$
$$
+ \sum_e{}' \int_e \{N\}_3\, p_3\psi_3\, dy + \sum_e{}' \int_e \{N\}_4\, p_4\psi_4\, dy \tag{6.13}
$$

with

$$[A] = \sum_e \iint_e [p\{N_x\}\{N_x\}^{\mathrm{T}} + p\{N_y\}\{N_y\}^{\mathrm{T}}$$
$$- qk_0^2\{N\}\{N\}^{\mathrm{T}}]\,dx\,dy \tag{6.14}$$

where the summation $\sum_e$ extends over all different elements, the summation $\sum_e'$ extends over the elements related to the boundaries Γ_1 to Γ_4, and $\{N\}_1$, $\{N\}_2$, $\{N\}_3$, and $\{N\}_4$ are the shape function vectors on Γ_1, Γ_2, Γ_3, and Γ_4, respectively. Using these shape function vectors, the field ψ on Γ_1 to Γ_4 may be discretized as

$$\psi_1 = \{N\}_1^{\mathrm{T}}\{\psi\}_1, \quad \psi_2 = \{N\}_2^{\mathrm{T}}\{\psi\}_2$$
$$\psi_3 = \{N\}_3^{\mathrm{T}}\{\psi\}_3, \quad \psi_4 = \{N\}_4^{\mathrm{T}}\{\psi\}_4 \tag{6.15}$$

where the components of the $\{\psi\}_1$ to $\{\psi\}_4$ vectors are the values of ψ at nodes on Γ_1 to Γ_4, respectively.

Equation (6.13) may be rewritten with the matrices in partitioned form

$$\begin{bmatrix} [A]_{00} & [A]_{01} & [A]_{02} & [A]_{03} & [A]_{04} \\ [A]_{10} & [A]_{11} & [A]_{12} & [A]_{13} & [A]_{14} \\ [A]_{20} & [A]_{21} & [A]_{22} & [A]_{23} & [A]_{24} \\ [A]_{30} & [A]_{31} & [A]_{32} & [A]_{33} & [A]_{34} \\ [A]_{40} & [A]_{41} & [A]_{42} & [A]_{43} & [A]_{44} \end{bmatrix} \begin{bmatrix} \{\phi\}_0 \\ \{\phi\}_1 \\ \{\phi\}_2 \\ \{\phi\}_3 \\ \{\phi\}_4 \end{bmatrix} = \begin{bmatrix} \{0\} \\ [B]_1\{\psi\}_1 \\ [B]_2\{\psi\}_2 \\ [B]_3\{\psi\}_3 \\ [B]_4\{\psi\}_4 \end{bmatrix} \tag{6.16}$$

with

$$[B]_1 = \sum_e'\int_e p_1\{N\}_1\{N\}_1^{\mathrm{T}}\,dx, \quad [B]_2 = \sum_e'\int_e p_2\{N\}_2\{N\}_2^{\mathrm{T}}\,dx$$
$$[B]_3 = \sum_e'\int_e p_3\{N\}_3\{N\}_3^{\mathrm{T}}\,dy, \quad [B]_4 = \sum_e'\int_e p_4\{N\}_4\{N\}_4^{\mathrm{T}}\,dy \tag{6.17}$$

where the components of the $\{\phi\}_0$ vector are the values of ϕ at nodes in the region Ω except boundaries Γ_1 to Γ_4, and $[A]_{00}$, $[A]_{01}$, $\cdots$, $[A]_{44}$ are the submatrices of $[A]$.

Considering the periodic boundary conditions, Eqs. (6.8) and (6.9), into Eq. (6.16), we obtain

$$\begin{bmatrix} [A]_{00} & [A]_{01} & [A]_{02} & [\tilde{A}]_{03} \\ [A]_{10} & [A]_{11} & [A]_{12} & [\tilde{A}]_{13} \\ [A]_{20} & [A]_{21} & [A]_{22} & [\tilde{A}]_{23} \\ [\tilde{A}]_{30} & [\tilde{A}]_{31} & [\tilde{A}]_{32} & [\tilde{A}]_{33} \end{bmatrix} \begin{bmatrix} \{\phi\}_0 \\ \{\phi\}_1 \\ \{\phi\}_2 \\ \{\phi\}_3 \end{bmatrix} = \begin{bmatrix} \{0\} \\ [B]_1\{\psi\}_1 \\ [B]_2\{\psi\}_2 \\ \{0\} \end{bmatrix} \tag{6.18}$$

with

$$[\tilde{A}]_{03} = [A]_{03} + \sigma[A]_{04} \tag{6.19a}$$

$$[\tilde{A}]_{13} = [A]_{13} + \sigma[A]_{14} \tag{6.19b}$$

$$[\tilde{A}]_{23} = [A]_{23} + \sigma[A]_{24} \tag{6.19c}$$

$$[\tilde{A}]_{30} = [A]_{30} + \sigma^*[A]_{40} \tag{6.19d}$$

$$[\tilde{A}]_{31} = [A]_{31} + \sigma^*[A]_{41} \tag{6.19e}$$

$$[\tilde{A}]_{32} = [A]_{32} + \sigma^*[A]_{42} \tag{6.19f}$$

$$[\tilde{A}]_{33} = [A]_{33} + [A]_{44} + \sigma[A]_{34} + \sigma^*[A]_{43} . \tag{6.19g}$$

6.2.3 Mode expansion

The internal fields in the regions $y \leq y_1$ and $y \geq y_2$ are represented by the sum of the normal modes as

$$\phi_i = \sum_{m=-\infty}^{\infty} [a_{mi} \exp(-j\kappa_{mi}y) + b_{mi} \exp(j\kappa_{mi}y)]f_{mi}(x) \tag{6.20}$$

$$\pm p_i \psi_i = \sum_{m=-\infty}^{\infty} j\kappa_{mi}[a_{mi} \exp(-j\kappa_{mi}y) - b_{mi} \exp(j\kappa_{mi}y)]g_{mi}(x)$$

$$\tag{6.21}$$

with

$$f_{mi}(x) = h_{mi}(x) \Big/ \sqrt{D_{mi}} \tag{6.22}$$

$$g_{mi}(x) = p_i h_{mi}(x) \Big/ \sqrt{D_{mi}} = p_i f_{mi}(x) \tag{6.23}$$

$$D_{mi} = \int_0^d p_i |h_{mi}(x)|^2 \, dx \tag{6.24}$$

$$h_{mi}(x) = \exp(-j\beta_m x) \tag{6.25}$$

$$\beta_m = n_1 k_0 \sin\theta + 2m\pi/d \tag{6.26}$$

$$\kappa_{mi} = \sqrt{n_i^2 k_0^2 - \beta_m^2}, \quad \text{Im}(\kappa_{mi}) \leq 0 \tag{6.27}$$

where a_{mi} and b_{mi} ($m = 0, \pm 1, \pm 2, \cdots$) are the amplitudes of the mth mode, the subscripts $i = 1$ and 2 designate the solutions in the regions $y \leq y_1$ and $y \geq y_2$, respectively, and the double signs $+$ and $-$ are for $i = 1$ and 2, respectively. The mode functions $f_{mi}(x)$ and $g_{mi}(x)$ satisfy the following orthonormalization relation:

$$\int_0^d f_{mi}^*(x)g_{m'i}(x)\,dx = \int_0^d g_{mi}^*(x)f_{m'i}(x)\,dx = \delta_{mm'} \tag{6.28}$$

where $\delta_{mm'}$ is the Kronecker delta.

6.2.4   Analytical approach

Assuming that the fundamental mode $(m = 0)$ of unit amplitude is incident from the negative y direction at an angle θ as shown in Fig. 6.1, the amplitudes of normal modes may be written as

$$a_{m1} = \begin{cases} 1 & \text{for } m = 0 \\ 0 & \text{for } m \neq 0 \end{cases} \tag{6.29a}$$

$$b_{m2} = 0. \tag{6.29b}$$

Considering Eq. (6.29) into Eqs. (6.20) and (6.21), from the orthonormalization relation, Eq. (6.28), we may express the field ϕ on the boundary Γ_i $(i = 1, 2)$ analytically as

$$\phi(x, y_i) = \delta_{i1} 2\exp(-j\kappa_{01}y_1)f_{01}(x)$$
$$- \sum_{m=-\infty}^{\infty}\left[\frac{f_{mi}(x)}{j\kappa_{mi}}\int_0^d g_{mi}^*(x')\psi(x', y_i)\,dx'\right]. \tag{6.30}$$

Equation (6.30) may be discretized as follows:

$$\{\phi\}_i = \delta_{i1}\{f\}_1 - [Z]_i\{\psi\}_i \tag{6.31}$$

with

$$\{f\}_1 = 2\exp(-j\kappa_{01}y_1)\{f_0\}_1 \tag{6.32}$$

$$[Z]_i = \sum_{m=-\infty}^{\infty}\left[\frac{\{f_m\}_i}{j\kappa_{mi}}\sum_e{}'\int_e g_{mi}^*(x')\{N(x', y_i)\}^{\mathrm{T}}\,dx'\right] \tag{6.33}$$

where the components of the $\{f_m\}_i$ vector are the values of $f_{mi}(x)$ at nodes on the boundary Γ_i and $\{N(x, y_i)\}$ is the shape function vector on Γ_i.

6.2.5 Combination of finite element and analytical relations

Substituting Eq. (6.31) into Eq. (6.18), we obtain the following final matrix equation:

$$
\begin{bmatrix}
[A]_{00} & [A]_{01} & [A]_{02} & [\tilde{A}]_{03} & [0] & [0] \\
[A]_{10} & [A]_{11} & [A]_{12} & [\tilde{A}]_{13} & -[B]_1 & [0] \\
[A]_{20} & [A]_{21} & [A]_{22} & [\tilde{A}]_{23} & [0] & -[B]_2 \\
[\tilde{A}]_{30} & [\tilde{A}]_{31} & [\tilde{A}]_{32} & [\tilde{A}]_{33} & [0] & [0] \\
[0] & [1] & [0] & [0] & [Z]_1 & [0] \\
[0] & [0] & [1] & [0] & [0] & [Z]_2
\end{bmatrix}
\begin{bmatrix}
\{\phi\}_0 \\
\{\phi\}_1 \\
\{\phi\}_2 \\
\{\phi\}_3 \\
\{\psi\}_1 \\
\{\psi\}_2
\end{bmatrix}
=
\begin{bmatrix}
\{0\} \\
\{0\} \\
\{0\} \\
\{0\} \\
\{f\}_1 \\
\{0\}
\end{bmatrix}
$$

$$(6.34)$$

The solutions of Eq. (6.34) allow the determination of the relative reflected and transmitted powers of the mth mode, ξ_m and η_m, as follows:

$$
\xi_m = \frac{\kappa_{m1}}{\kappa_{01}} \left| \delta_{m0} \exp(-j\kappa_{01}y_1) - \frac{1}{j\kappa_{m1}} \int_0^d g_{m1}^*(x)\psi(x,y_1)\,dx \right|^2
$$

$$(6.35)$$

$$
\eta_m = \frac{1}{\kappa_{01}\kappa_{m2}} \left| \int_0^d g_{m2}^*(x)\psi(x,y_2)\,dx \right|^2 .
$$

$$(6.36)$$

6.3 Surface Impedance Approximation

We consider a metallic grating with period d in the x direction as shown in Fig. 6.2(a), where the boundary corresponding to the grating surface is Γ_5.

Regarding a metallic grating as a lossy dielectric grating and applying the FEM to the region Ω in Fig. 6.2(b), the diffraction characteristics can be evaluated from Eq. (6.34).

When the surface impedance approximation is used, the FEM is applied to only the region Ω_s in Fig. 6.2(c), and in place of Eq. (6.34), the following matrix equation may be derived:

$$
\begin{bmatrix}
[A]_{00} & [A]_{01} & [\tilde{A}]_{03} & [A]_{05} & [0] \\
[A]_{10} & [A]_{11} & [\tilde{A}]_{13} & [A]_{15} & -[B]_1 \\
[\tilde{A}]_{30} & [\tilde{A}]_{31} & [\tilde{A}]_{33} & [\tilde{A}]_{35} & [0] \\
[A]_{50} & [A]_{51} & [\tilde{A}]_{53} & [A]_{55} & [0] \\
[0] & [1] & [0] & [0] & [Z]_1
\end{bmatrix}
\begin{bmatrix}
\{\phi\}_0 \\
\{\phi\}_1 \\
\{\phi\}_3 \\
\{\phi\}_5 \\
\{\psi\}_1
\end{bmatrix}
=
\begin{bmatrix}
\{0\} \\
\{0\} \\
\{0\} \\
\sum_e{}' \int_e \{N\}_5\, p_5\psi_5\, d\Gamma \\
\{f\}_1
\end{bmatrix} .
$$

$$(6.37)$$

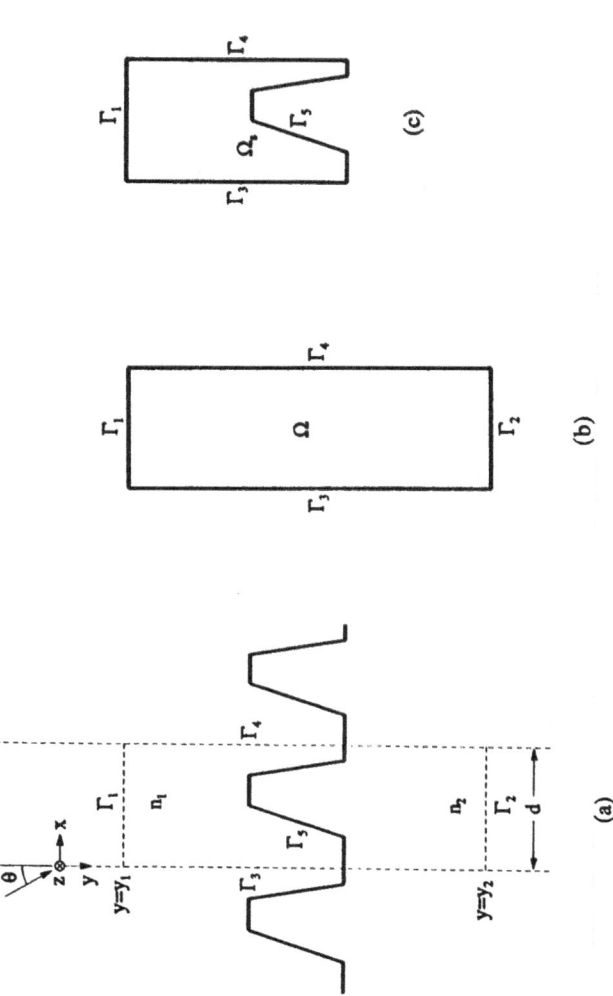

Figure 6.2: Metallic grating. (a) Grating structure. (b) Region Ω for the two-medium boundary-value problem. (c) Region Ω_s for the surface impedance approximation method.

For the boundary condition on Γ_5, the surface impedance Z_s is introduced as

$$Z_s = Z_0/n_m \tag{6.38}$$

where Z_0 is the impedance of free space (see Eq. (2.6)) and n_m is the refractive index of metal.

When the continuity of electromagnetic fields is considered, the following relations between ϕ and ψ on the boundary Γ_5 hold:

$$p_5\psi_5 = \begin{cases} -(j\omega\mu_0/Z_s)\phi_5 & \text{for TE modes} \\ -j\omega\varepsilon_0 Z_s\phi_5 & \text{for TM modes.} \end{cases} \tag{6.39}$$

Using the shape function vector $\{N\}_5$ on Γ_5, the field ϕ_5 on Γ_5 may be discretized as

$$\phi_5 = \{N\}_5^{\mathrm{T}}\{\phi\}_5 . \tag{6.40}$$

Substituting Eqs. (6.39) and (6.40) into Eq. (6.37), we obtain the following final matrix equation:

$$\begin{bmatrix} [A]_{00} & [A]_{01} & [\tilde{A}]_{03} & [A]_{05} & [0] \\ [A]_{10} & [A]_{11} & [\tilde{A}]_{13} & [A]_{15} & -[B]_1 \\ [\tilde{A}]_{30} & [\tilde{A}]_{31} & [\tilde{A}]_{33} & [\tilde{A}]_{35} & [0] \\ [A]_{50} & [A]_{51} & [\tilde{A}]_{53} & [\tilde{A}]_{55} & [0] \\ [0] & [1] & [0] & [0] & [Z]_1 \end{bmatrix} \begin{bmatrix} \{\phi\}_0 \\ \{\phi\}_1 \\ \{\phi\}_3 \\ \{\phi\}_5 \\ \{\psi\}_1 \end{bmatrix} = \begin{bmatrix} \{0\} \\ \{0\} \\ \{0\} \\ \{0\} \\ \{f\}_1 \end{bmatrix} \tag{6.41}$$

with

$$[\tilde{A}]_{55} = [A]_{55} + [Z]_5 \tag{6.42}$$

$$[Z]_5 = \begin{cases} \sum_e' \int_e (j\omega\mu_0/Z_s)\{N\}_5\{N\}_5^{\mathrm{T}}\, d\Gamma & \text{for TE modes} \\ \sum_e' \int_e j\omega\varepsilon_0 Z_s\{N\}_5\{N\}_5^{\mathrm{T}}\, d\Gamma & \text{for TM modes.} \end{cases} \tag{6.43}$$

Metallic gratings are often analyzed as perfectly conducting gratings. The FEM for perfectly conducting gratings is almost identical to the FEM using the surface impedance approximation, except that the boundary condition for the boundary Γ_5 is given as follows:

$$\phi_5 = 0 \quad \text{for TE modes} \tag{6.44}$$

$$\psi_5 = 0 \quad \text{for TM modes.} \tag{6.45}$$

The final matrix equation may be written as

$$\begin{bmatrix} [A]_{00} & [A]_{01} & [\tilde{A}]_{03} & [0] \\ [A]_{10} & [A]_{11} & [\tilde{A}]_{13} & -[B]_1 \\ [\tilde{A}]_{30} & [\tilde{A}]_{31} & [\tilde{A}]_{33} & [0] \\ [0] & [1] & [0] & [Z]_1 \end{bmatrix} \begin{bmatrix} \{\phi\}_0 \\ \{\phi\}_1 \\ \{\phi\}_3 \\ \{\psi\}_1 \end{bmatrix} = \begin{bmatrix} \{0\} \\ \{0\} \\ \{0\} \\ \{f\}_1 \end{bmatrix} \tag{6.46}$$

for the TE modes, and for the TM modes the matrix $[\tilde{A}]_{55}$ in Eq. (6.41) should be replaced by the matrix $[A]_{55}$.

6.4 Dielectric Gratings

First, we consider the sinusoidal, trapezoidal, and rectangular dielectric gratings as shown in Fig. 6.3. Figure 6.4 shows an example of the element division. The parameters used in the calculations are $d = 0.6\lambda$, $n_1 = 1.0$, and $n_2 = 2.0$.

Figures 6.5 and 6.6 show the diffraction characteristics for the TE and TM mode incidences, respectively. The results for the sinusoidal and rectangular gratings for the TE mode incidence agree well with those of the unimoment method.[8]

Next, a holographic grating[9] as shown in Fig. 6.7(a) is investigated, where $d = 0.458$ μm, $n_1 = 1.0$, and $n_2 = 1.64$. Figure 6.7(b) shows the element division for the holographic grating.

Figure 6.8 shows the wavelength dependence of the diffraction efficiency (namely, the relative transmitted power of the -1-order mode) at the fixed incident angle $\theta = 36.86°$. The solid lines, the dashed lines, and the dots show the results of the FEM, the results of the coupled-mode theory (CMT),[9] and the experimental results,[9] respectively. Under the above parameters, the 0- and -1-order reflected and transmitted modes can propagate in the structure through the axis of abscissa and the -2- and $+1$-order transmitted modes can propagate in the ranges $\lambda < 0.5129$ μm and $\lambda < 0.4764$ μm, respectively. Accordingly, Wood's anomalies are expected to appear at the cutoff wavelengths $\lambda = 0.4764$ μm and $\lambda = 0.5129$ μm. In the case of the TE mode incidence, Wood's anomalies can be seen clearly. In the case of the TM mode incidence, on the other hand, the anomalies cannot be seen very clearly.

Figure 6.9 shows the incident angle dependence of the diffraction efficiency at the fixed wavelength $\lambda = 0.458$ μm. The 0- and -1-order modes can propagate in the same manner as in Fig. 6.8, and the -2- and $+1$-order transmitted modes can propagate in the ranges $\theta > 21.1°$ and $\theta < 39.8°$, respectively.

Figure 6.10 shows the wavelength dependence of diffraction efficiency under the Bragg condition given by

$$2d\sin\theta = m\lambda \qquad \text{for } -m\text{th-order mode.} \qquad (6.47)$$

Under this condition the wavelength and the incident angle change simultaneously. It is reported that, if the wavelength and the incident angle satisfy the Bragg condition, the diffraction efficiency is generally (but not always) maximized for the $-m$-th-order mode. In Figs. 6.8 and 6.9, the Bragg wavelength and the incident angle satisfying Eq. (6.47) of $m = 1$ correspond to

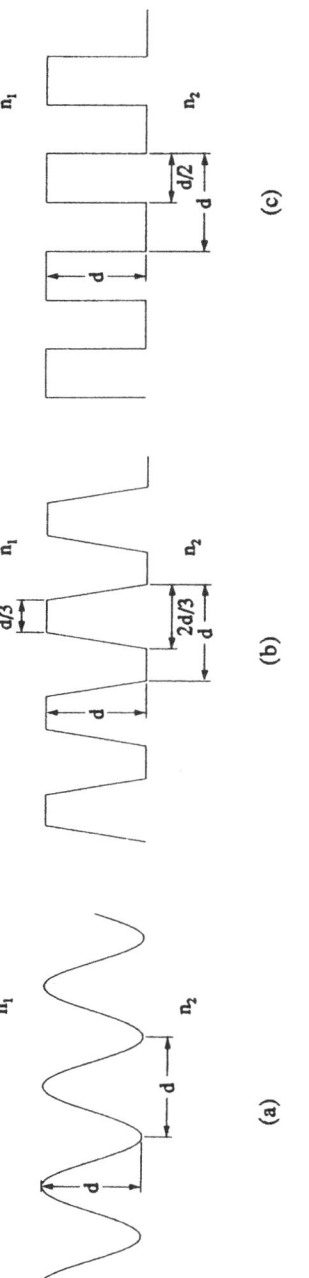

Figure 6.3:   Dielectric grating profiles.   (a) Sinusoidal grating. (b) Trapezoidal grating. (c) Rectangular grating.

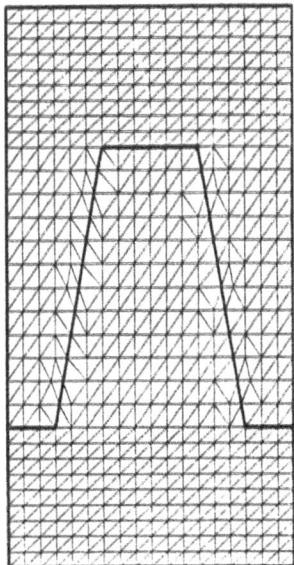

Figure 6.4: Element division of a trapezoidal grating (1080 elements and 2257 nodes are used).

$\lambda = 0.5495$ μm and $\theta = 30°$, respectively, and the diffraction efficiency of the -1-order mode undoubtedly takes a maximum near these values in the case of the TE mode incidence. Under the condition of Eq. (6.47) of $m = 1$, the 0- and -1-order modes can propagate in the same manner as in Figs. 6.8 and 6.9, and the -2- and $+1$-order transmitted modes can propagate in the range $\lambda < 0.5007$ μm (or $\theta < 33.1°$). In Figs. 6.8 to 6.10, the results of the FEM agree well with the experimental data[9] but not with those of the CMT[9] in the case of the TE mode incidence. In the case of the TM mode incidence, the results of the FEM agree well with those of the CMT[9] but not with the experimental data.[9] The reason for the discrepancy between the cases of the TE and TM mode incidences is unknown, but the results of the FEM in Figs. 6.8 to 6.10 agree excellently with those of the boundary element method.[6] In the above calculations, the energy error is less than 0.01%.

6.5 Metallic Gratings

We consider the sinusoidal, triangular, and rectangular metallic gratings as shown in Fig. 6.11.

Figure 6.12 shows the diffraction characteristics for the sinusoidal gold

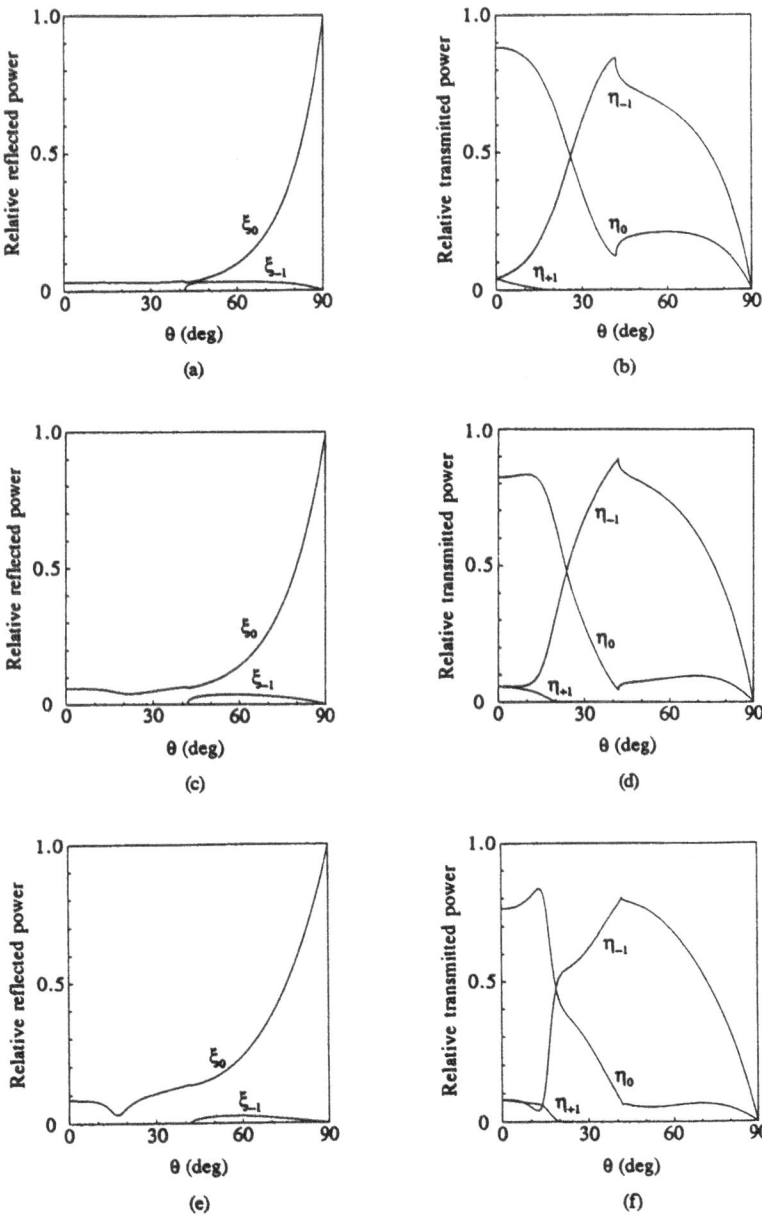

Figure 6.5: Diffraction characteristics for the TE mode incidence. (a) Reflected power for a sinusoidal grating. (b) Transmitted power for a sinusoidal grating. (c) Reflected power for a trapezoidal grating. (d) Transmitted power for a trapezoidal grating. (e) Reflected power for a rectangular grating. (f) Transmitted power for a rectangular grating.

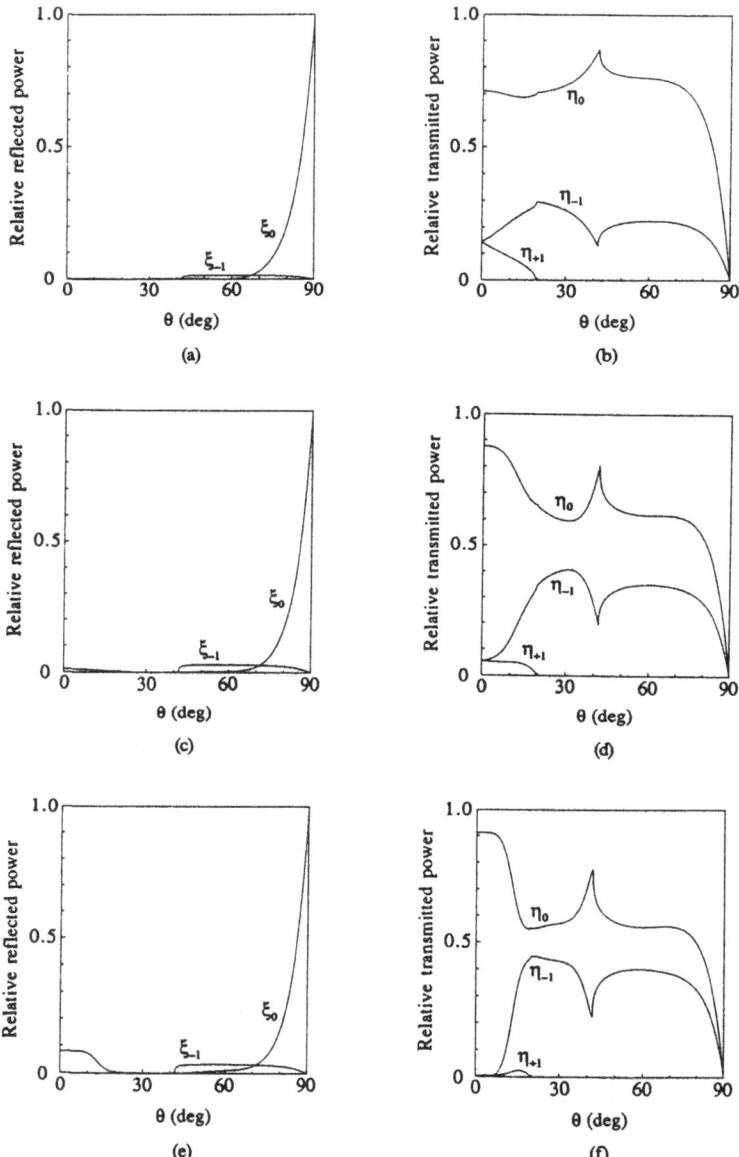

Figure 6.6: Diffraction characteristics for the TM mode incidence. (a) Reflected power for a sinusoidal grating. (b) Transmitted power for a sinusoidal grating. (c) Reflected power for a trapezoidal grating. (d) Transmitted power for a trapezoidal grating. (e) Reflected power for a rectangular grating. (f) Transmitted power for a rectangular grating.

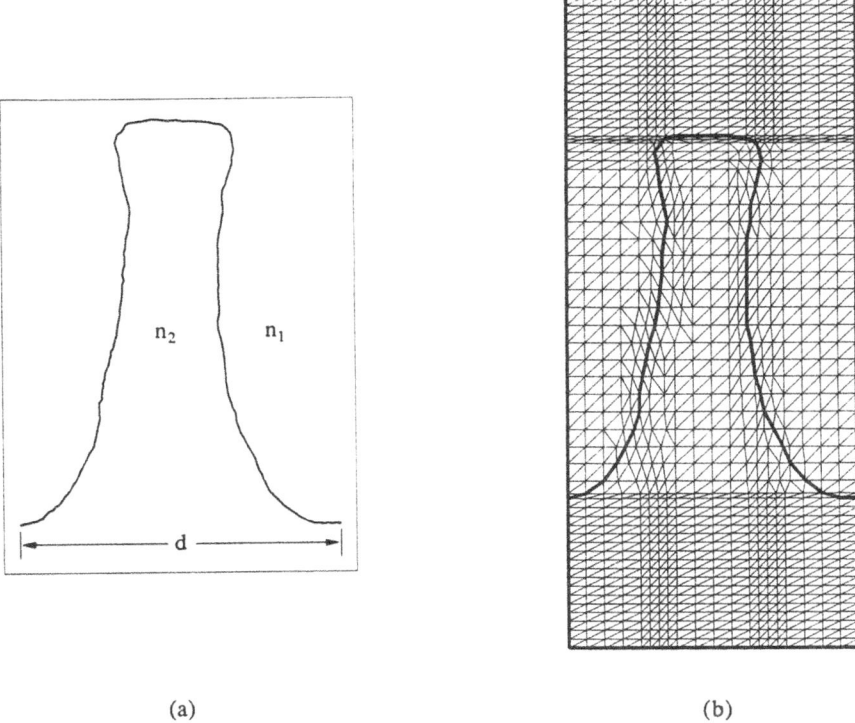

(a)   (b)

Figure 6.7: Holographic grating. (a) Grating profile. (b) Element division (2360 elements and 4879 nodes are used).

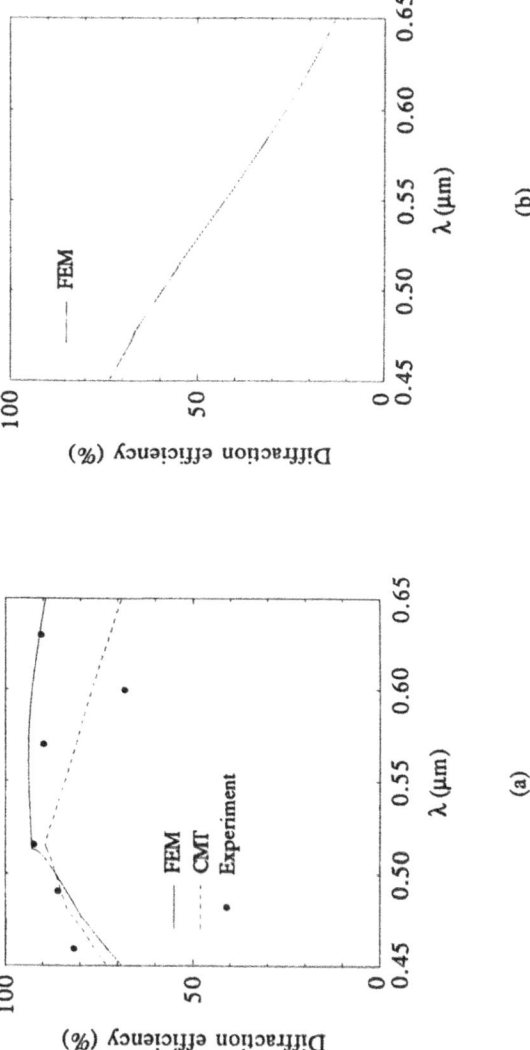

Figure 6.8:   Wavelength dependence of diffraction efficiency at the fixed incident angle $\theta = 36.86°$. (a) TE mode. (b) TM mode.

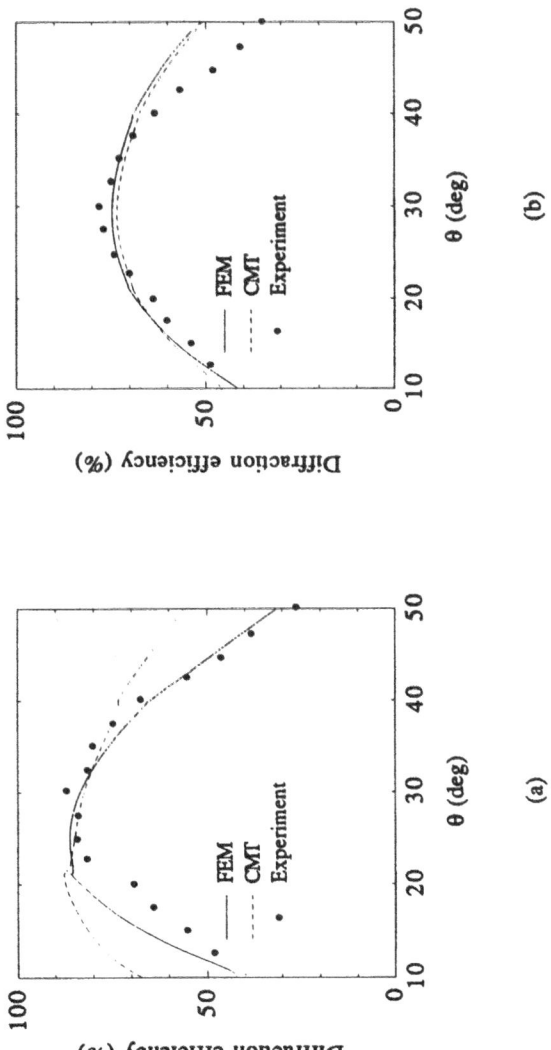

Figure 6.9: Incident angle dependence of diffraction efficiency at the fixed wavelength $\lambda = 0.458\ \mu$m. (a) TE mode. (b) TM mode.

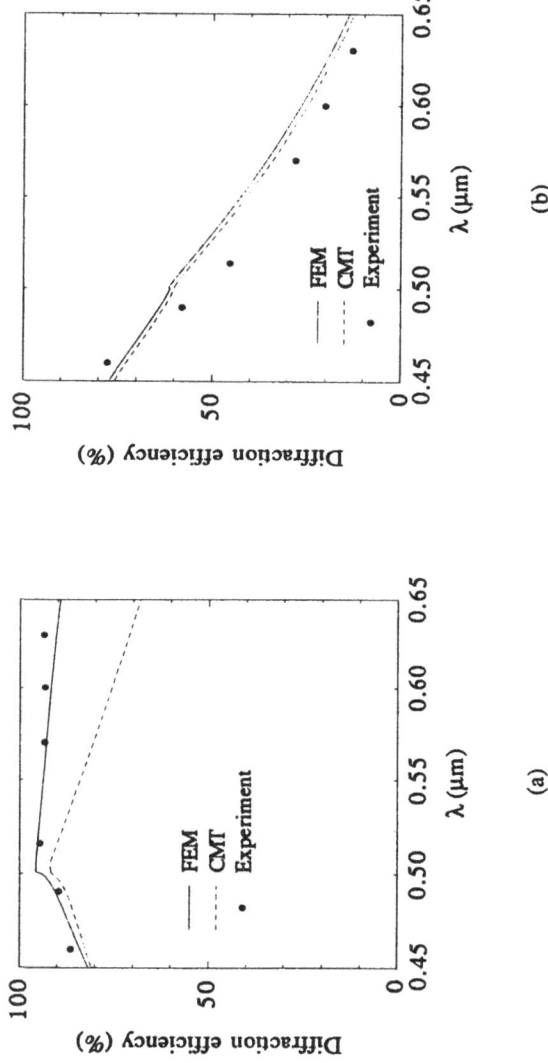

Figure 6.10:   Wavelength dependence of diffraction efficiency under the Bragg condition. (a) TE mode. (b) TM mode.

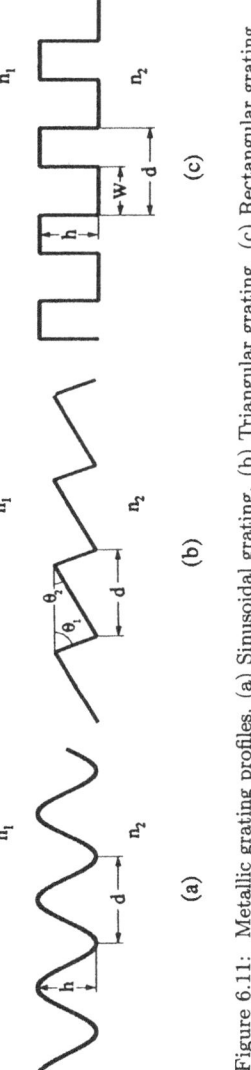

Figure 6.11: Metallic grating profiles. (a) Sinusoidal grating. (b) Triangular grating. (c) Rectangular grating.

grating, where ξ_t is the total reflected power and

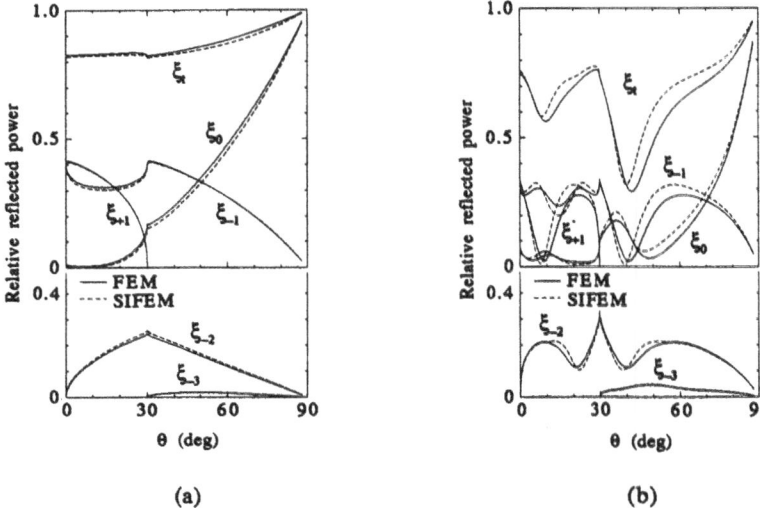

(a)   (b)

Figure 6.12: Diffraction characteristics for a sinusoidal gold grating. (a) TE mode. (b) TM mode.

$$\lambda = 0.550 \ \mu m \,, \quad d = 1.1 \ \mu m \,, \quad h = 0.11 \ \mu m$$
$$n_1 = 1.0 \,, \quad n_m = 0.32 - j2.32 \,.$$

The solid lines indicate the results of the FEM, where the diffraction problem of metallic gratings is exactly analyzed as a two-medium boundary-value problem.[6] On the other hand, the dashed lines indicate the results of a combination of the surface impedance approximation and finite element methods (SIFEM). The results of the SIFEM agree well with those of the integral equation method using the surface impedance approximation.[10),11)] Figure 6.13 shows the diffraction characteristics for the perfectly conducting grating with sinusoidal profile, where grating parameters are the same as those in Fig. 6.12.

Figure 6.14 shows the diffraction characteristics for the sinusoidal silver grating, where

$$\lambda = 0.476 \ \mu m \,, \quad d = 1.205 \ \mu m \,, \quad h = 0.2 \ \mu m$$
$$n_1 = 1.0 \,, \quad n_m = 0.052 - j2.65 \,.$$

The results of the FEM agree well with those of the extended boundary condition method.[12]

Figure 6.15 shows the diffraction characteristics for the triangular silver grating, where

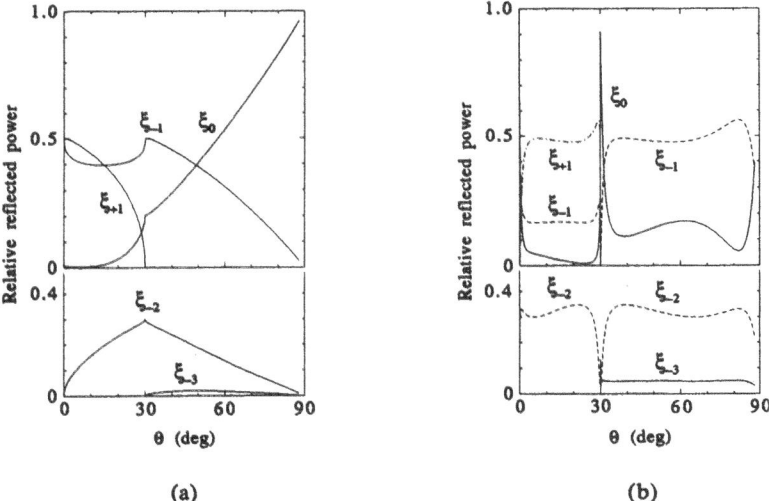

Figure 6.13: Diffraction characteristics for a perfectly conducting grating with sinusoidal profile. (a) TE mode. (b) TM mode.

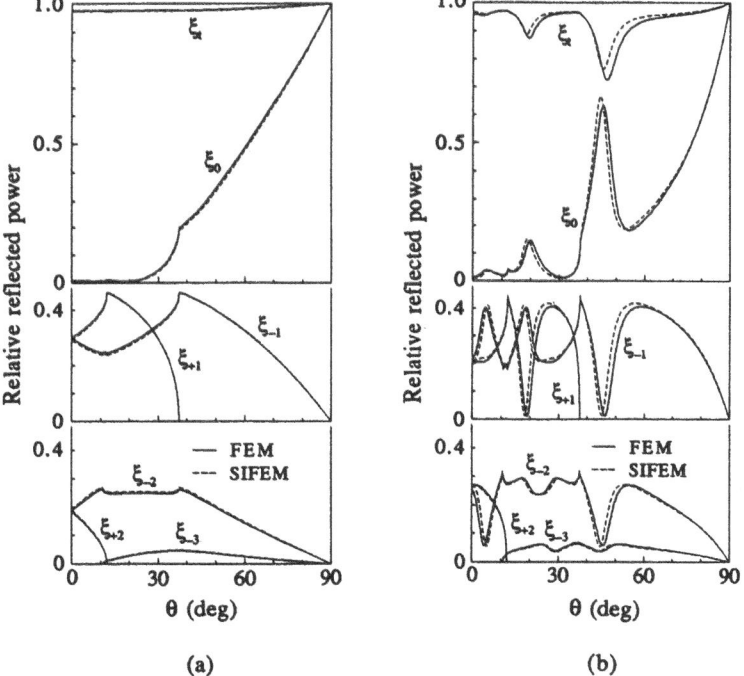

Figure 6.14: Diffraction characteristics for a sinusoidal silver grating. (a) TE mode. (b) TM mode.

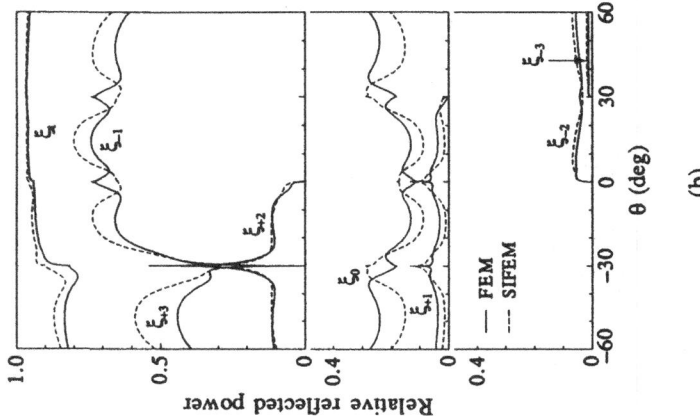

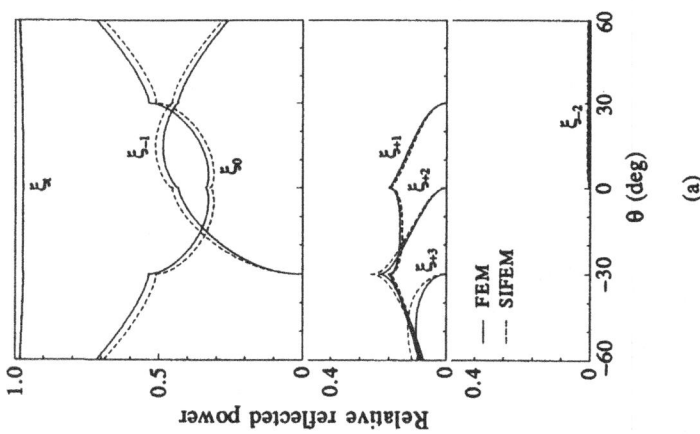

Figure 6.15:   Diffraction characteristics for a triangular silver grating. (a) TE mode. (b) TM mode.

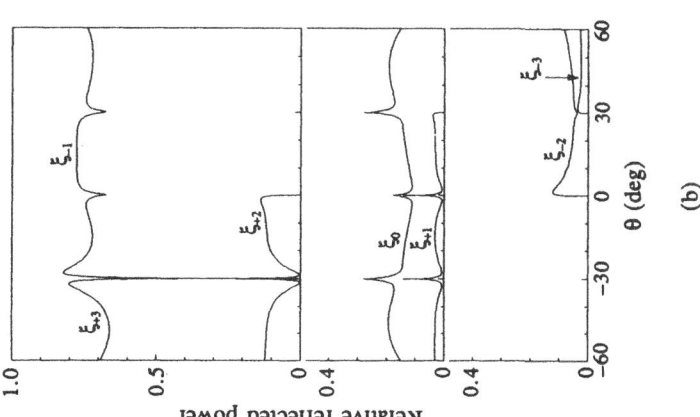

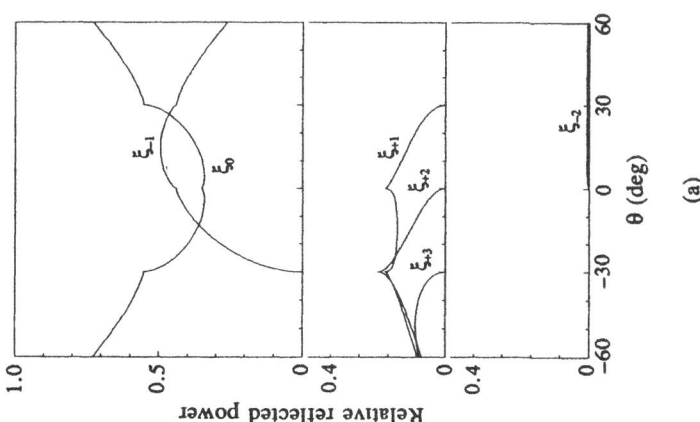

Figure 6.16: Diffraction characteristics for a perfectly conducting grating with triangular profile. (a) TE mode. (b) TM mode.

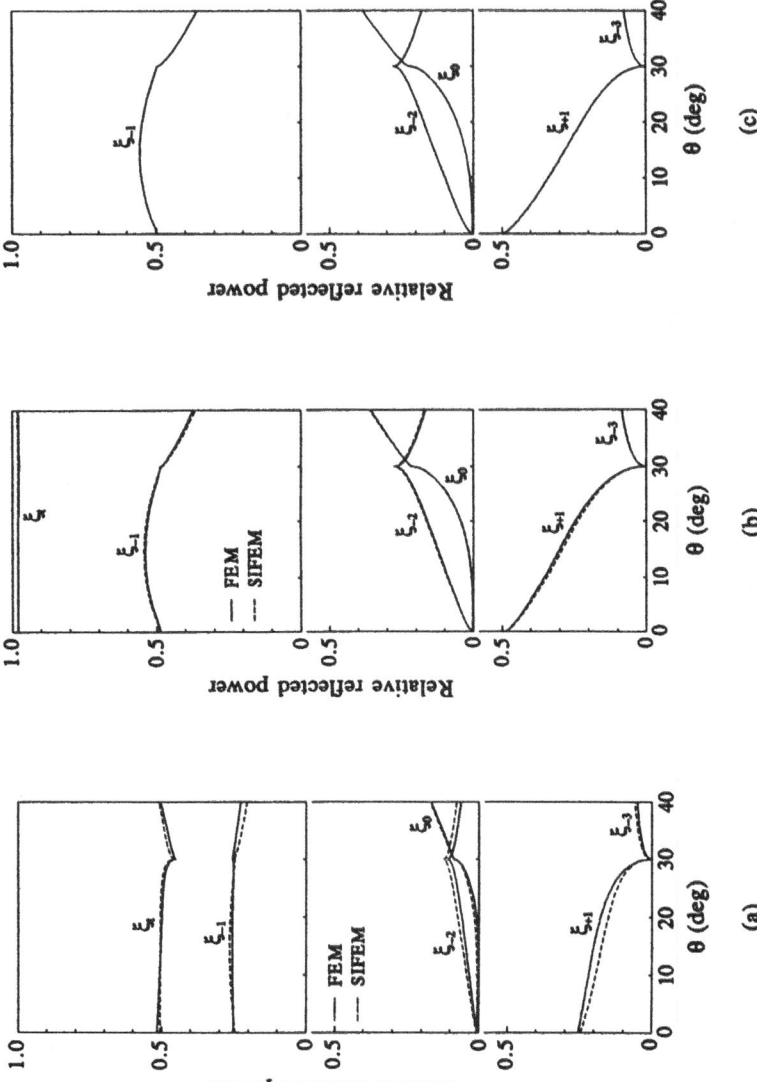

Figure 6.17:   Diffraction characteristics for the TE mode of a rectangular grating.   (a) $\lambda = 0.5$ μm.   (b) $\lambda = 1.0$ μm. (c) Perfectly conducting grating.

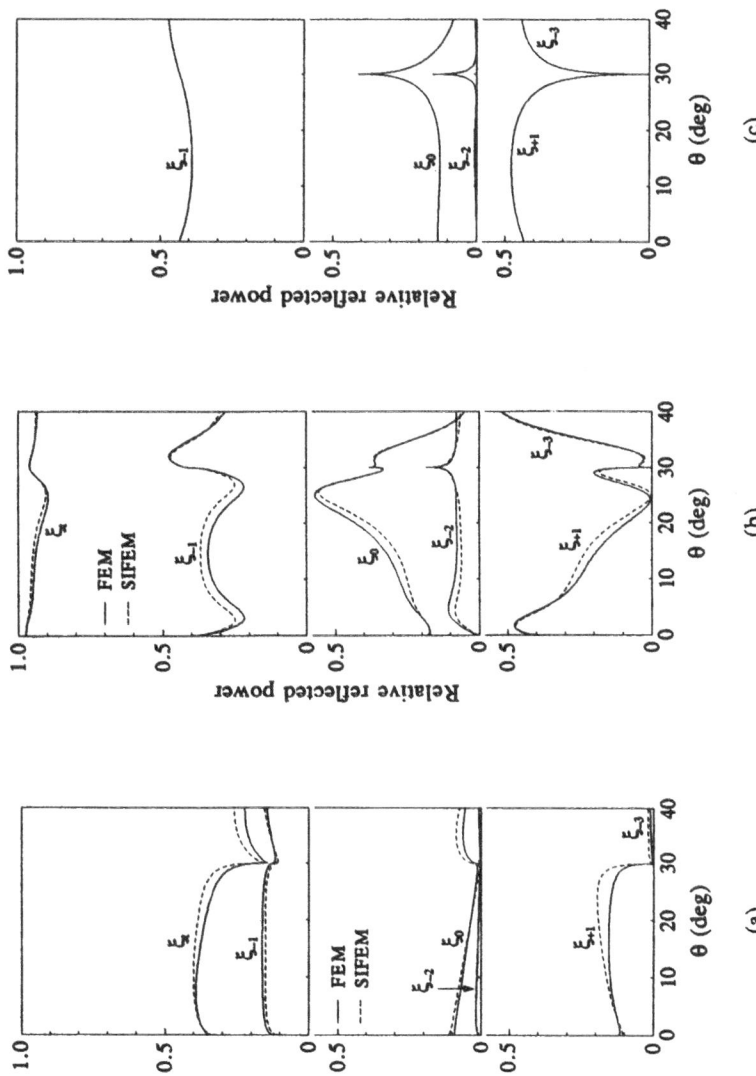

Figure 6.18: Diffraction characteristics for the TM mode of a rectangular grating. (a) $\lambda = 0.5 \ \mu m$. (b) $\lambda = 1.0 \ \mu m$. (c) Perfectly conducting grating.

$$\lambda = 0.476 \ \mu\text{m}, \quad d = 0.952 \ \mu\text{m}, \quad \theta_1 = 80°, \quad \theta_2 = 10°$$
$$n_1 = 1.0, \quad n_m = 0.052 - j2.65.$$

Figure 6.16 shows the diffraction characteristics for the perfectly conducting grating with triangular profile, where grating parameters are the same as those in Fig. 6.15, and the results obtained agree well with those of the mode matching method.[13]

Figures 6.17 and 6.18 show the diffraction characteristics of the TE and TM modes for the rectangular gold grating, respectively, where

$$d = 2\lambda, \quad h = 0.3\lambda, \quad W = 0.5d$$
$$n_1 = 1.0$$
$$n_m = \begin{cases} 0.803 - j1.818 & \text{for } \lambda = 0.5 \ \mu\text{m} \\ 0.220 - j6.710 & \text{for } \lambda = 1.0 \ \mu\text{m}. \end{cases}$$

The results obtained agree well with those of the CMT.[14]

From Figs. 6.12 to 6.18 it is confirmed that in the case of the TE mode incidence, the results of the SIFEM agree approximately with those of the FEM and that the diffraction characteristics of perfectly conducting gratings resemble those of lossy metallic gratings. In the case of the TM mode incidence, on the other hand, the results of the SIFEM deviate from those of the FEM, and the diffraction characteristics of perfectly conducting gratings are significantly different from those of lossy metallic gratings. Especially for the TM mode incidence, the metallic gratings should be analyzed with metallic loss taken into account.

References

1) R. Petit (ed.), *Electromagnetic Theory of Gratings*, Springer-Verlag, New York, 1980.

2) M. K. Moaveni, H. A. Kalhor, and S. Shammas, "Application of finite-elements to the analysis of diffraction gratings", *Int. Jour. Electron.*, Vol. 40, No. 3, pp. 225–236, March 1976.

3) M. K. Moaveni, "Analysis of transmission gratings by the method of finite elements", *IEE Proc.*, Vol. 126, No. 1, pp. 35–40, Jan. 1979.

4) M. Koshiba, H. Nemoto, and M. Suzuki, "Application of finite-element method to scattering from periodic structures", *Trans. Inst. Electron. Commun. Eng. Japan*, Vol. E66, No. 11, pp. 684–685, Nov. 1983.

5) M. K. Moaveni, A. A. Rizvi, and B. A. Kamran, "Plane-wave scattering by gratings of conducting cylinders embedded in an inhomogeneous and lossy dielectric", *Jour. Opt. Soc. Am. A*, Vol. 5, No. 6, pp. 834–842, June 1988.

6) Y. Nakata and M. Koshiba, "Boundary-element analysis of plane-wave diffraction from groove-type dielectric and metallic gratings", *Jour. Opt. Soc. Am. A*, Vol. 7, No. 8, pp. 1494–1502, Aug. 1990.

7) S. D. Gedney and R. Mittra, "Analysis of the electromagnetic scattering by thick gratings using a combined FEM/MM solution", *IEEE Trans. Antennas Propagat.*, Vol. 39, No. 11, pp. 1605–1614, Nov. 1991.

8) D. E. Tremain and K. K. Mei, "Application of the unimoment method to scattering from periodic dielectric structures", *Jour. Opt. Soc. Am.*, Vol. 68, No. 6, pp. 775–783, June 1978.

9) M. G. Moharam, T. K. Gaylord, G. T. Sincerbox, H. Werlich, and B. Yung, "Diffraction characteristics of photoresist surface-relief gratings", *Appl. Opt.*, Vol. 23, No. 18, pp. 3214–3220, Sept. 1984.

10) R. A. Depine, "Perfectly conducting diffraction grating formalisms extended to good conductors via the surface impedance boundary condition", *Appl. Opt.*, Vol. 26, No. 12, pp. 2348–2354, June 1987.

11) R. A. Depine and J. M. Simon, "Comparison between the differential and integral methods used to solve the grating problem in the $H_\parallel$ case", *Jour. Opt. Soc. Am. A*, Vol. 4, No. 5, pp. 834–838, May 1987.

12) S. L. Chuang and J. A. Kong, "Wave scattering and guidance by dielectric waveguides with periodic surfaces", *Jour. Opt. Soc. Am.*, Vol. 73, No. 5, pp. 669–679, May 1983.

13) K. Yasuura and Y. Okuno, "Numerical analysis of diffraction from a grating by the mode-matching method with a smoothing procedure", *Jour. Opt. Soc. Am.*, Vol. 72, No. 7, pp. 847–852, July 1982.

14) M. G. Moharam and T. K. Gaylord, "Rigorous coupled-wave analysis of metallic surface-relief gratings", *Jour. Opt. Soc. Am. A*, Vol. 3, No. 11, pp. 1780–1787, Nov. 1986.

Chapter 7

OPTICAL WAVEGUIDE DISCONTINUITIES

7.1 Introduction

Optical waveguide discontinuities play an important role in designing optical components. Various theoretical methods for the solution of dielectric slab waveguide discontinuities have been developed. Although these methods are very useful for a step discontinuity or a cascade of steps, it seems to be difficult to apply them to arbitrarily-shaped discontinuities. Recently, the integral equation method (IEM), the boundary element method (BEM), and the FEM[1)−19)] have been presented for the solution of arbitrarily-shaped discontinuities. In the IEM, however, only the weakly guiding structure is considered. Also, the BEM[4)] cannot be effectively applied to a problem involving inhomogeneous media. The FEM is very useful for the arbitrarily-shaped discontinuities including inhomogeneous media. However, it is in general difficult to apply the FEM to discontinuities in an open-type dielectric waveguide.

This chapter presents a numerical approach based on a combination of finite and boundary elements for the solution of discontinuities in an open dielectric slab waveguide.[10)−12),17)−19)] The discontinuity region is divided into two regions. One is a finite region with arbitrary inhomogeneities; the other is a semi-infinite and homogeneous region. The FEM and the BEM are applied to the former and the latter region, respectively. The finite element can be combined with the boundary element on the common nodes because these two methods are discretized in the same way. Also, analytical solutions in which both the guided and radiated modes are taken into account are used for uniform waveguide regions connected to discontinuities.

7.2   Method of Analysis

7.2.1   Basic equations

We consider an asymmetric planar waveguide as shown in Fig. 7.1. The boundaries $\Gamma_{+\infty}$ and $\Gamma_{-\infty}$ are placed at infinity ($y = \pm\infty$) and the boundary $\Gamma_i = \Gamma_{ci} + \Gamma_{fi} + \Gamma_{si}$ ($i = 1, 2$) connects the discontinuity region to the uniform waveguide i, where t_i is the thickness of waveguide i and n_c, n_f, and n_s ($n_f > n_s > n_c$) are the refractive indices of the cover, film, and substrate, respectively. The region Ω_f surrounded by the boundary $\Gamma_f = \Gamma_{f1} + \Gamma_{f2} + \Gamma_{c3} + \Gamma_{s3}$ completely encloses the discontinuity region, the region Ω_c is surrounded by the boundaries $\Gamma_c = \Gamma_{c1} + \Gamma_{c2} + \Gamma_{c3}$ and $\Gamma_{+\infty}$, and the region Ω_s is surrounded by the boundaries $\Gamma_s = \Gamma_{s1} + \Gamma_{s2} + \Gamma_{s3}$ and $\Gamma_{-\infty}$.

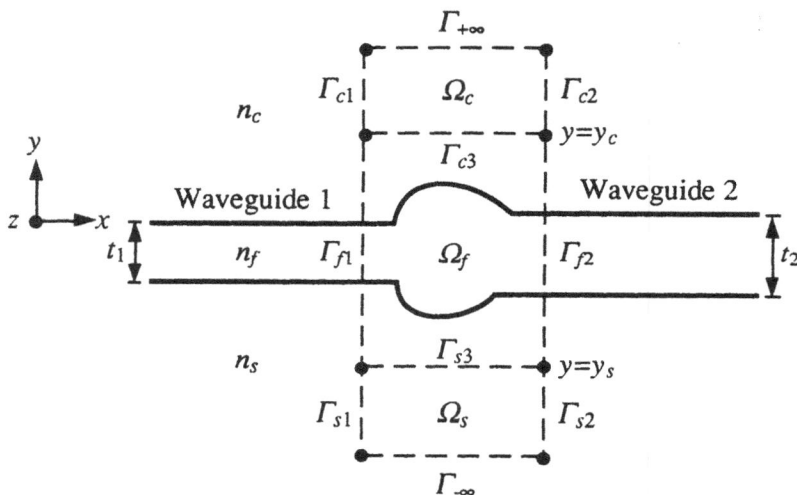

Figure 7.1: Optical waveguide discontinuity.

Assuming that there is no variation of fields and refractive indices in the horizontal transverse z direction, we consider the following basic equation for harmonic wave propagation in the x direction:

$$\nabla_t \cdot (p\nabla_t \phi) + q k_0^2 \phi = 0 \tag{7.1}$$

with

$$\nabla_t = i_x \partial/\partial x + i_y \partial/\partial y \tag{7.2}$$

where i_x and i_y are the unit vectors in the x and y directions, respectively, and the field ϕ and the coefficients p, q are written as follows:

(1) TE modes:

$$\phi = E_z \tag{7.3}$$

$$p = 1 \tag{7.4a}$$

$$q = n^2 . \tag{7.4b}$$

(2) TM modes:

$$\phi = H_z \tag{7.5}$$

$$p = 1/n^2 \tag{7.6a}$$

$$q = 1 . \tag{7.6b}$$

We define ψ on Γ_1, Γ_2, Γ_{c3}, and Γ_{s3} as follows:

$$\psi_1 = -\partial\phi/\partial x \quad \text{on } \Gamma_1 \tag{7.7a}$$

$$\psi_2 = \partial\phi/\partial x \quad \text{on } \Gamma_2 \tag{7.7b}$$

$$\psi_{c3} = -\partial\phi/\partial y \quad \text{on } \Gamma_{c3} \tag{7.7c}$$

$$\psi_{s3} = \partial\phi/\partial y \quad \text{on } \Gamma_{s3} . \tag{7.7d}$$

The functional for Eq. (7.1) is given by

$$
\begin{aligned}
F = & \iint_{\Omega_f} [p(\nabla_t \phi)^* \cdot (\nabla_t \phi) - q k_0^2 \phi^* \phi] \, dx \, dy \\
& - \int_{\Gamma_{f1}} p_{f1} \phi_{f1}^* \psi_{f1} \, d\Gamma - \int_{\Gamma_{f2}} p_{f2} \phi_{f2}^* \psi_{f2} \, d\Gamma \\
& + \int_{\Gamma_{c3}} p_{c3} \phi_{c3}^* \psi_{c3} \, d\Gamma + \int_{\Gamma_{s3}} p_{s3} \phi_{s3}^* \psi_{s3} \, d\Gamma .
\end{aligned}
\tag{7.8}
$$

7.2.2 Finite element approach

Dividing the region Ω_f into a number of quadratic triangular elements (see Fig. 1.3(b)), we expand the field ϕ in each element as

$$\phi = \{N\}^{\mathrm{T}} \{\phi\}_e . \tag{7.9}$$

Substituting Eq. (7.9) into Eq. (7.8), from the variational principle we obtain the following global matrix equation:

$$
\begin{aligned}
[A]\{\phi\} = & \sum_e{}' \int_e \{N\}_{f1} \, p_{f1} \psi_{f1} \, dy + \sum_e{}' \int_e \{N\}_{f2} \, p_{f2} \psi_{f2} \, dy \\
& - \sum_e{}' \int_e \{N\}_{c3} \, p_{c3} \psi_{c3} \, dx - \sum_e{}' \int_e \{N\}_{s3} \, p_{s3} \psi_{s3} \, dx \quad (7.10)
\end{aligned}
$$

with

$$[A] \;=\; \sum_e \iint_e \left[p\{N_x\}\{N_x\}^{\mathrm{T}} + p\{N_y\}\{N_y\}^{\mathrm{T}} \right.$$
$$\left. - qk_0^2\{N\}\{N\}^{\mathrm{T}} \right] dx\, dy \qquad (7.11)$$

where $\{N\}_{f1}$, $\{N\}_{f2}$, $\{N\}_{c3}$, and $\{N\}_{s3}$ are the shape function vectors on Γ_{f1}, Γ_{f2}, Γ_{c3}, and Γ_{s3}, respectively. Using these shape function vectors, the field ψ on Γ_{f1}, Γ_{f2}, Γ_{c3}, and Γ_{s3} may be discretized as

$$\psi_{f1} = \{N\}_{f1}^{\mathrm{T}}\{\psi\}_{f1}\,, \quad \psi_{f2} = \{N\}_{f2}^{\mathrm{T}}\{\psi\}_{f2}$$
$$\psi_{c3} = \{N\}_{c3}^{\mathrm{T}}\{\psi\}_{c3}\,, \quad \psi_{s3} = \{N\}_{s3}^{\mathrm{T}}\{\psi\}_{s3} \qquad (7.12)$$

where the components of the $\{\psi\}_{f1}$, $\{\psi\}_{f2}$, $\{\psi\}_{c3}$, and $\{\psi\}_{s3}$ vectors are the values of ψ at nodes on Γ_{f1}, Γ_{f2}, Γ_{c3}, and Γ_{s3}, respectively.

Eliminating internal variables, namely the nodes in Ω_f except Γ_f, we obtain the following small-sized matrix equation:

$$[A]_f \begin{bmatrix} \{\phi\}_{f1} \\ \{\phi\}_{f2} \\ \{\phi\}_{c3} \\ \{\phi\}_{s3} \end{bmatrix} = [B]_f \begin{bmatrix} \{\psi\}_{f1} \\ \{\psi\}_{f2} \\ \{\psi\}_{c3} \\ \{\psi\}_{s3} \end{bmatrix} \qquad (7.13)$$

with

$$[B]_f \;=\; [\, [B]_{f1}\; [B]_{f2}\; [B]_{c3}\; [B]_{s3}\,] \qquad (7.14)$$

$$[B]_{f1} \;=\; \sum_e{}' \int_e p_{f1}\{N\}_{f1}\{N\}_{f1}^{\mathrm{T}}\, dy$$

$$[B]_{f2} \;=\; \sum_e{}' \int_e p_{f2}\{N\}_{f2}\{N\}_{f2}^{\mathrm{T}}\, dy$$

$$[B]_{c3} \;=\; -\sum_e{}' \int_e p_{c3}\{N\}_{c3}\{N\}_{c3}^{\mathrm{T}}\, dx$$

$$[B]_{s3} \;=\; -\sum_e{}' \int_e p_{s3}\{N\}_{s3}\{N\}_{s3}^{\mathrm{T}}\, dx \qquad (7.15)$$

where $[A]_f$ is the modified matrix of $[A]$ due to the elimination of internal nodes.

7.2.3 Boundary element approach

Applying the BEM (see Subsection 1.6.4) with quadratic line elements (see Fig. 1.13(b)) to the regions Ω_c and Ω_s, and considering the radiation condition on $\Gamma_{\pm\infty}$, we obtain the following matrix equation:

$$[H]_c\{\phi\}_c = [G]_c\{\psi\}_c \quad \text{for } \Omega_c \qquad (7.16\text{a})$$
$$[H]_s\{\phi\}_s = [G]_s\{\psi\}_s \quad \text{for } \Omega_s \qquad (7.16\text{b})$$

where the (i,j) components of the matrices $[H]$ and $[G]$ are given by

$$h_{ij} = \delta_{ij}\frac{\theta_i}{2\pi} + \sum_e \oint_e \frac{\partial G_i}{\partial n} N_j \, d\Gamma \qquad (7.17)$$

$$g_{ij} = \sum_e \oint_e G_i N_j \, d\Gamma . \qquad (7.18)$$

Here θ_i is the interior angle at the node i and equal to π if the boundary is smooth there, G_i is the Green's function, N_j is the shape function for the BEM, and $\oint$ denotes Cauchy's principle value of integration. Calculation formulas for Eqs. (7.17) and (7.18) are presented in Subsection 1.6.4.

7.2.4 Mode expansion

The internal fields in the uniform waveguide i $(i = 1, 2)$ are represented by the sum of the normal modes as

$$\phi_i = \sum_{m=0}^{M_i-1} [a_{mi}\exp(-j\beta_{mi}x) + b_{mi}\exp(j\beta_{mi}x)]f_{mi}(y)$$

$$+ \sum_{r=0}^{2}\int_{\rho_{r1}}^{\rho_{r2}} \{a_{ri}(\rho)\exp[-j\beta(\rho)x]$$

$$+ b_{ri}(\rho)\exp[j\beta(\rho)x]\}f_{ri}(y,\rho)\,d\rho \qquad (7.19)$$

$$\pm p_i(y)\psi_i = \sum_{m=0}^{M_i-1} j\beta_{mi}[a_{mi}\exp(-j\beta_{mi}x) - b_{mi}\exp(j\beta_{mi}x)]g_{mi}(y)$$

$$+ \sum_{r=0}^{2}\int_{\rho_{r1}}^{\rho_{r2}} j\beta(\rho)\{a_{ri}(\rho)\exp[-j\beta(\rho)x]$$

$$- b_{ri}(\rho)\exp[j\beta(\rho)x]\}g_{ri}(y,\rho)\,d\rho \qquad (7.20)$$

with

$$\rho_{r1} = \begin{cases} 0 & \text{for } r = 0 \\ k_0\sqrt{n_s^2 - n_c^2} & \text{for } r = 1, 2 \end{cases} \qquad (7.21\text{a})$$

$$\rho_{r2} = \begin{cases} k_0\sqrt{n_s^2 - n_c^2} & \text{for } r = 0 \\ \infty & \text{for } r = 1, 2 \end{cases} \qquad (7.21\text{b})$$

where M_i is the number of guided modes in waveguide i, β_{mi} and a_{mi}, b_{mi} ($m = 0, 1, 2, \cdots$) are the propagation constant and amplitudes of the mth guided mode, respectively, $\beta(\rho)$ and $a_{ri}(\rho)$, $b_{ri}(\rho)$ are the propagation constant and amplitudes of radiation modes, respectively, ρ is the wavenumber of the radiation mode in the y direction in the substrate region, and the double signs $+$ and $-$ are for $i = 1$ and 2, respectively. The mode functions $f_{mi}(y)$, $g_{mi}(y)$, $f_{ri}(y, \rho)$, and $g_{ri}(y, \rho)$ satisfy the following orthonormalization relation:

$$\int_{-\infty}^{\infty} f_{mi}^*(y) g_{m'i}(y)\, dy = \int_{-\infty}^{\infty} g_{mi}^*(y) f_{m'i}(y)\, dy = \delta_{mm'} \qquad (7.22\text{a})$$

$$\int_{-\infty}^{\infty} f_{ri}^*(y, \rho) g_{r'i}(y, \rho')\, dy = \int_{-\infty}^{\infty} g_{ri}^*(y, \rho) f_{r'i}(y, \rho')\, dy = \delta_{rr'} \delta(\rho - \rho')$$

$$(7.22\text{b})$$

where $\delta_{mm'}$, $\delta_{rr'}$ are the Kronecker deltas and $\delta(\rho - \rho')$ is the Dirac delta fuction. The mode functions and propagation constants are summarized as follows (for simplicity the subscript i is omitted; the origin of the y axis is taken as the interface between cover and film):

(1) Guided modes:

$$f_m(y) = \frac{1}{\sqrt{D_m}} h_m(y) \qquad (7.23)$$

$$g_m(y) = \frac{p(y)}{\sqrt{D_m}} h_m(y) = p(y) f_m(y) \qquad (7.24)$$

$$p(y) = \begin{cases} p_c & \text{for } y \geq 0 \text{ (cover)} \\ p_f & \text{for } -t \leq y \leq 0 \text{ (film)} \\ p_s & \text{for } y \leq -t \text{ (substrate)} \end{cases} \qquad (7.25)$$

$$D_m = \int_{-\infty}^{\infty} p(y) |h_m(y)|^2\, dy \qquad (7.26)$$

$$h_m = \begin{cases} \exp(-\alpha_{cm} y) & \text{for } y \geq 0 \\ \cos \kappa_{fm} y - \dfrac{p_c \alpha_{cm}}{p_f \kappa_{fm}} \sin \kappa_{fm} y & \\ & \text{for } -t \leq y \leq 0 \quad (7.27) \\ \left(\cos \kappa_{fm} t + \dfrac{p_c \alpha_{cm}}{p_f \kappa_{fm}} \sin \kappa_{fm} t \right) & \\ \times \exp[\alpha_{sm}(y + t)] & \text{for } y \leq -t \end{cases}$$

$$\alpha_{cm} = \sqrt{\beta_m^2 - n_c^2 k_0^2} \qquad (7.28)$$

$$\kappa_{fm} = \sqrt{n_f^2 k_0^2 - \beta_m^2} \qquad (7.29)$$

$$\alpha_{sm} = \sqrt{\beta_m^2 - n_s^2 k_0^2} \tag{7.30}$$

where p_c, p_f, and p_s are defined as

$$p_c = p_f = p_s = 1 \tag{7.31a}$$

for the TE modes, and

$$p_c = 1/n_c^2, \quad p_f = 1/n_f^2, \quad p_s = 1/n_s^2 \tag{7.31b}$$

for the TM modes. The characteristic equation for the propagation constant β_m is given by

$$\tan(\kappa_{fm}t) = \frac{p_f \kappa_{fm}(p_c \alpha_{cm} + p_s \alpha_{sm})}{p_f^2 \kappa_{fm}^2 - p_c p_s \alpha_{cm} \alpha_{sm}} . \tag{7.32}$$

(2) Radiation modes:

$$f_r(y, \rho) = \frac{1}{\sqrt{D_r(\rho)}} h_r(y, \rho) \tag{7.33}$$

$$g_r(y, \rho) = \frac{p(y)}{\sqrt{D_r(\rho)}} h_r(y, \rho) = p(y) f_r(y, \rho) \tag{7.34}$$

$$D_r(\rho)\delta(\rho - \rho') = \int_{-\infty}^{\infty} p(y) h_r^*(y, \rho) h_r(y, \rho') \, dy \tag{7.35}$$

where $h_r(y, \rho)$ is given by

$$h_r(y, \rho) = \begin{cases} \exp(-\alpha_c y) & \text{for } y \geq 0 \\ \cos \kappa_f y - \dfrac{p_c \alpha_c}{p_f \kappa_f} \sin \kappa_f y & \text{for } -t \leq y \leq 0 \\ \left(\cos \kappa_f t + \dfrac{p_c \alpha_c}{p_f \kappa_f} \sin \kappa_f t \right) \cos[\rho(y+t)] & \\ + \left(\dfrac{p_f \kappa_f}{p_s \rho} \sin \kappa_f t - \dfrac{p_c \alpha_c}{p_s \rho} \cos \kappa_f t \right) & \\ \times \sin[\rho(y+t)] & \text{for } y \leq -t \end{cases} \tag{7.36}$$

with

$$\alpha_c = \sqrt{(n_s^2 - n_c^2)k_0^2 - \rho^2} \tag{7.37}$$

$$\kappa_f = \sqrt{(n_f^2 - n_s^2)k_0^2 + \rho^2} \tag{7.38}$$

$$\beta(\rho) = \sqrt{n_s^2 k_0^2 - \rho^2} \tag{7.39}$$

for the substrate radiation modes ($r = 0$), and

$$
h_r(y, \rho) = \begin{cases}
\cos \kappa_c y - \dfrac{p_f \kappa_f}{p_c \kappa_c} c_r \sin \kappa_c y & \text{for } y \geq 0 \\
\cos \kappa_f y - c_r \sin \kappa_f y & \text{for } -t \leq y \leq 0 \\
(\cos \kappa_f t + c_r \sin \kappa_f t) \cos[\rho(y + t)] & \\
\quad + \dfrac{p_f \kappa_f}{p_s \rho}(\sin \kappa_f t - c_r \cos \kappa_f t) & \\
\quad \times \sin[\rho(y + t)] & \text{for } y \leq -t
\end{cases}
\tag{7.40}
$$

with

$$
\kappa_c = \sqrt{(n_c^2 - n_s^2)k_0^2 + \rho^2} \tag{7.41}
$$

$$
\beta(\rho) = \begin{cases}
\sqrt{n_s^2 k_0^2 - \rho^2} & \text{for propagating part} \\
-j\sqrt{\rho^2 - n_s^2 k_0^2} & \text{for nonpropagating part}
\end{cases}
\tag{7.42}
$$

$$
c_1 = \left(\frac{p_s^2 \rho^2 \cos^2 \kappa_f t + p_f^2 \kappa_f^2 \sin^2 \kappa_f t + p_c p_s \kappa_c \rho}{p_f^2 \kappa_f^2 \cos^2 \kappa_f t + p_s^2 \rho^2 \sin^2 \kappa_f t + p_f^2 \kappa_f^2 p_s \rho / p_c \kappa_c} \right)^{1/2}
\tag{7.43}
$$

$$
c_2 = -c_1 \tag{7.44}
$$

for the substrate-cover radiation modes ($r = 1, 2$).

7.2.5 Analytical approach

Assuming that the fundamental mode ($m = 0$) of unit amplitude is incident from the left side of waveguide 1 in Fig. 7.1, the amplitudes of normal modes may be written as

$$
a_{m1} = \begin{cases} 1 & \text{for } m = 0 \\ 0 & \text{for } m \neq 0 \end{cases}
\tag{7.45a}
$$

$$
b_{m2} = a_{r1}(\rho) = b_{r2}(\rho) = 0 \,. \tag{7.45b}
$$

Considering Eq. (7.45) into Eqs. (7.19) and (7.20), and using the ortho-normalization relation, Eq. (7.22), we may express the field ϕ on the boundary Γ_i ($i = 1, 2$) analytically as

$$
\begin{aligned}
\phi(x_i, y) = {} & \delta_{i1} 2 \exp(-j\beta_{01} x_1) f_{01}(y) \\
& - \sum_{m=0}^{M_i - 1} \frac{f_{mi}(y)}{j\beta_{mi}} \int_{-\infty}^{\infty} g_{mi}^*(y') \psi(x_i, y') \, dy' \\
& - \sum_{r=0}^{2} \int_{\rho_{r1}}^{\rho_{r2}} \frac{f_{ri}(y, \rho)}{j\beta(\rho)} \int_{-\infty}^{\infty} g_{ri}^*(y', \rho) \psi(x_i, y') \, dy' \, d\rho \,.
\end{aligned}
\tag{7.46}
$$

Equation (7.46) may be discretized as follows:

$$\{\phi\}_i = \delta_{i1}\{f\}_1 - [Z]_i\{\psi\}_i \tag{7.47}$$

with

$$\{f\}_1 = 2\exp(-j\beta_{01}x_1)\{f_0\}_1 \tag{7.48}$$

$$[Z]_i = \sum_{m=0}^{M_i-1} \frac{\{f_m\}_i}{j\beta_{mi}} \sum_e{}' \int_e g_{mi}^*(y')\{N(x_i,y')\}^T \, dy'$$

$$+ \sum_{r=0}^{2} \int_{\rho_{r1}}^{\rho_{r2}} \frac{\{f_r(\rho)\}_i}{j\beta(\rho)} \sum_e{}' \int_e g_{ri}^*(y',\rho)$$

$$\times \{N(x_i,y')\}^T \, dy' \, d\rho \tag{7.49}$$

where the components of the $\{f_m\}_i$ and $\{f_r(\rho)\}_i$ vectors are the values of $f_{mi}(y)$ and $f_{ri}(y,\rho)$ at nodes on the boundary Γ_i, respectively, and $\{N(x_i,y)\}$ is the shape function vector on Γ_i.

7.2.6 Combination of finite element, boundary element, and analytical relations

From Eqs. (7.13), (7.16), and (7.47) we obtain the following final matrix equation:

$$\begin{bmatrix} [\tilde{A}]_f & \vdots & -[\tilde{B}]_f \\ [\tilde{H}]_c & \vdots & -[\tilde{G}]_c \\ [\tilde{H}]_s & \vdots & -[\tilde{G}]_s \\ \cdots & \cdots & \cdots & \cdots & \cdots & \cdots & \cdots & \cdots \\ [1] & [0] & [0] & [0] & \vdots & [Z]_1 & [0] & [0] & [0] \\ [0] & [1] & [0] & [0] & \vdots & [0] & [Z]_2 & [0] & [0] \end{bmatrix} \begin{bmatrix} \{\phi\}_1 \\ \{\phi\}_2 \\ \{\phi\}_c \\ \{\phi\}_s \\ \cdots \\ \{\psi\}_1 \\ \{\psi\}_2 \\ \{\psi\}_c \\ \{\psi\}_s \end{bmatrix} = \begin{bmatrix} \{0\} \\ \{0\} \\ \{0\} \\ \cdots \\ \{f\}_1 \\ \{0\} \end{bmatrix}$$

$$(7.50)$$

where the columns of the matrices $[\tilde{A}]_f$ and $[\tilde{B}]_f$ corresponding to the nodes on Γ_f are the same as those of the matrices $[A]_f$ and $[B]_f$, respectively, and the others are zero. The matrices $[\tilde{H}]_c$, $[\tilde{H}]_s$, $[\tilde{G}]_c$, and $[\tilde{G}]_s$ are generated in a similar way from the matrices $[H]_c$, $[H]_s$, $[G]_c$, and $[G]_s$, respectively.

The solutions of Eq. (7.50) allow the determination of the relative reflected and transmitted powers of the mth mode, ξ_m and η_m, and the relative radiated power in the waveguide i, ζ_i, as follows:

$$\xi_m = \frac{\beta_{m1}}{\beta_{01}} \left| \delta_{0m}\exp(-j\beta_{01}x_1) - \frac{1}{j\beta_{m1}} \int_{-\infty}^{\infty} g_{m1}^*(y)\psi(x_1,y)\, dy \right|^2$$

$$(7.51)$$

$$\eta_m = \frac{1}{\beta_{01}\beta_{m2}} \left| \int_{-\infty}^{\infty} g_{m2}^*(y)\psi(x_2,y)\,dy \right|^2 \tag{7.52}$$

$$\zeta_i = \frac{1}{\beta_{01}} \sum_{r=0}^{2} \int_{\rho_{r1}}^{\rho_{r2}'} \frac{1}{\beta(\rho)} \left| \int_{-\infty}^{\infty} g_{ri}^*(y,\rho)\psi(x_i,y)\,dy \right|^2 d\rho \tag{7.53}$$

with

$$\rho_{r2}' = \begin{cases} k_0\sqrt{n_s^2 - n_c^2} & \text{for } r = 0 \\ k_0 n_s & \text{for } r = 1,2 \, . \end{cases} \tag{7.54}$$

7.3 Discontinuities in an Asymmetric Planar Waveguide

For numerical computation, introducing a parameter D_c, we divide the integrals with respect to y in the boundary element approach for Ω_c into two parts, namely, those in $y_c \leq y \leq D_c$ and those in $D_c \leq y < \infty$. Similarly, for Ω_s, $-\infty < y \leq -D_s$ and $-D_s \leq y \leq y_s$. Also, we divide the integrals with respect to y in the analytical approach into three parts, namely, those in $-\infty < y \leq -D_s$, those in $-D_s \leq y \leq D_c$, and those in $D_c \leq y < \infty$. The integrals in $y_c \leq y \leq D_c$ and $-D_s \leq y \leq y_s$ for the boundary element approach, and in $-D_s \leq y \leq D_c$ for the analytical approach are calculated analytically, but the others are neglected by choosing the values of D_c and D_s adequately. Furthermore, introducing a parameter a, we divide the integrals with respect to ρ in the analytical relation into three parts, namely, those in $0 \leq \rho \leq n_s k_0$ (propagating part), those in $n_s k_0 \leq \rho \leq a n_s k_0$ (nonpropagating part), and those in $a n_s k_0 \leq \rho < \infty$ (nonpropagating part), where the first two parts are calculated numerically and the last part is neglected by choosing the value of a ($a > 1$) adequately. A double-exponential formula is used for the numerical integration over ρ.

We consider three kinds of step discontinuities as shown in Fig. 7.2, where $\lambda = 0.6328$ μm, $t_1 = 0.3$ μm or 0.4 μm, $n_c = 1$, $n_f = 1.61$, $n_s = 1.515$, and the fundamental TE or TM mode incidence is assumed. Convergence of the solution should be checked on values of a, D_c, and D_s. We investigate the convergence for step A with TE mode incidence and $t_1 = 0.3$ μm.

Table 7.1 shows the variation of solutions with values of a, where ζ and P_t represent total radiated power, $\zeta_1 + \zeta_2$, and total power, $\xi_0 + \eta_0 + \zeta$, respectively. Since the results for $a = 2$ are almost the same as those for $a = 4$, we use $a = 2$.

Appropriate values of D_c and D_s are dependent on the spread of the guided mode over the y direction. Tables 7.2 and 7.3 show the variation of solutions with some values of D_c and D_s, respectively, which are chosen

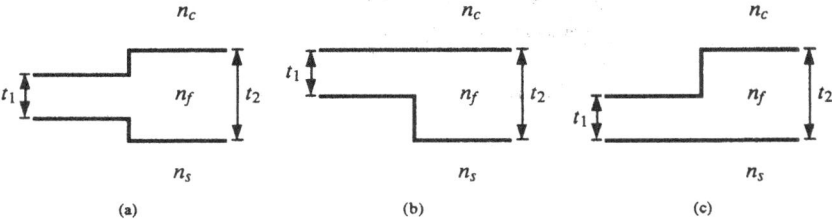

Figure 7.2: Step discontinuities. (a) Step A. (b) Step B. (c) Step C.

Table 7.1: Variation of solutions with values of a.

t_2/t_1	a	ξ_0	η_0	ζ	P_t
1.5	2	0.00003	0.93541	0.06455	0.99999
	4	0.00003	0.93540	0.06455	0.99999
2.0	2	0.00007	0.88892	0.11099	0.99998
	4	0.00007	0.88892	0.11099	0.99998

$$D_c = 2\lambda + (t_2 - t_1)/2 \,, \quad D_s = 4\lambda + (t_2 + t_1)/2 \,.$$

Table 7.2: Variation of solutions with values of D_c.

t_2/t_1	D_c	ξ_0	η_0	ζ	P_t
1.5	$\lambda + 0.25t_1$	0.00003	0.93541	0.06454	0.99998
	$3\lambda + 0.25t_1$	0.00003	0.93541	0.06455	0.99999
2.0	$\lambda + 0.5t_1$	0.00007	0.88892	0.11098	0.99997
	$3\lambda + 0.5t_1$	0.00007	0.88892	0.11099	0.99998

$$D_s = 4\lambda + (t_2 + t_1)/2 \,, \quad a = 2 \,.$$

Table 7.3: Variation of solutions with values of D_s.

t_2/t_1	D_s	ξ_0	η_0	ζ	P_t
1.5	$3\lambda + 1.25t_1$	0.00003	0.93539	0.06449	0.99991
	$5\lambda + 1.25t_1$	0.00003	0.93541	0.06453	0.99997
2.0	$3\lambda + 1.5t_1$	0.00007	0.88891	0.11094	0.99992
	$5\lambda + 1.5t_1$	0.00007	0.88892	0.11098	0.99997

$$D_c = 2\lambda + (t_2 - t_1)/2 \,, \quad a = 2 \,.$$

in view of a field distribution of the fundamental guided mode shown in Fig. 7.3. From these tables, it is reasonable to decide $D_c = \lambda + (t_2 - t_1)/2$ and $D_s = 4\lambda + (t_2 + t_1)/2$.

Figures 7.4 and 7.5 show the transmitted and radiated powers of steps A, B, and C for the TE and TM mode incidences, respectively. The results of the radiated power of step A agree well with those of the mode matching method (MMM).[20] In the case of the TE mode incidence of $t_1 = 0.4$ μm, the second-order mode ($m = 1$) exists in waveguide 2 for $t_2/t_1 \geq 1.97$, and especially for step C it is strongly excited. Also, we notice that the radiated power is large in order of steps B, A, and C except for the TE mode incidence of $t_1 = 0.4$ μm.

7.4   Discontinuities in a Symmetric Planar Waveguide

A combination of the finite element and boundary element methods (FBEM) can also be applied to discontinuities in a symmetric planar waveguide. The FBEM for the discontinuities in a symmetric waveguide ($n_c = n_s$) is almost identical to the FBEM for those in an asymmetric waveguide ($n_c \neq n_s$), except that the functions $h_m(y)$ (see Eq. (7.27)) and $h_r(y)$ (see Eqs. (7.36) and (7.40)) constructing the mode functions are rewritten as follows (the origin of the y axis is taken as the plane of symmetry):

(1) Even guided modes ($m = 0, 2, 4, \cdots$):

$$h_m(y) = \begin{cases} \cos \kappa_{fm} \dfrac{t}{2} \exp\left[-\alpha_{sm}\left(|y| - \dfrac{t}{2}\right)\right] & \text{for } |y| \geq t/2 \\ \cos \kappa_{fm} y & \text{for } |y| \leq t/2. \end{cases}$$

$$(7.55)$$

(2) Odd guided modes ($m = 1, 3, 5, \cdots$):

$$h_m(y) = \begin{cases} \pm \sin \kappa_{fm} \dfrac{t}{2} \exp\left[-\alpha_{sm}\left(|y| - \dfrac{t}{2}\right)\right] & \text{for } |y| \geq t/2 \\ \sin \kappa_{fm} y & \text{for } |y| \leq t/2. \end{cases}$$

$$(7.56)$$

(3) Even radiation modes ($r = 1$):

$$h_1(y, \rho) = \begin{cases} \cos \kappa_f \dfrac{t}{2} \cos \rho \left(|y| - \dfrac{t}{2}\right) \\ \quad - \dfrac{p_f \kappa_f}{p_s \rho} \sin \kappa_f \dfrac{t}{2} \sin \rho \left(|y| - \dfrac{t}{2}\right) & \text{for } |y| \geq t/2 \\ \cos \kappa_f y & \text{for } |y| \leq t/2. \end{cases}$$

$$(7.57)$$

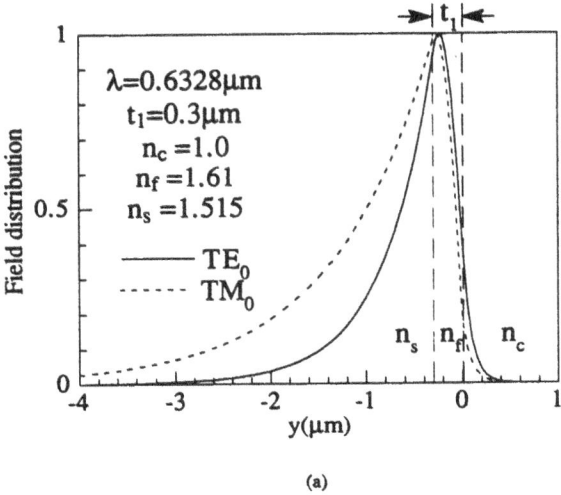

(a)

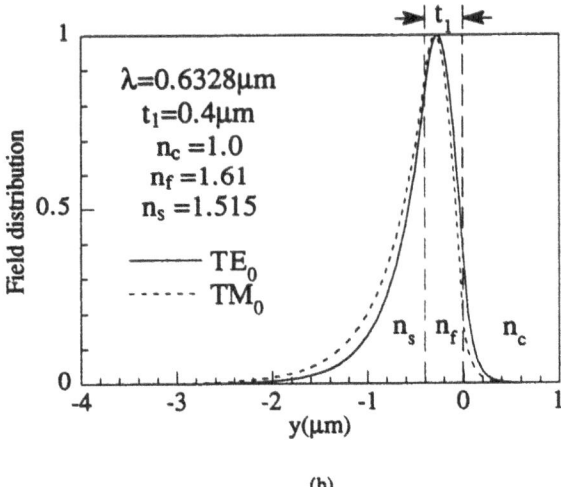

(b)

Figure 7.3: Field distribution of the fundamental guided mode. (a) $t_1 = 0.3$ μm. (b) $t_1 = 0.4$ μm.

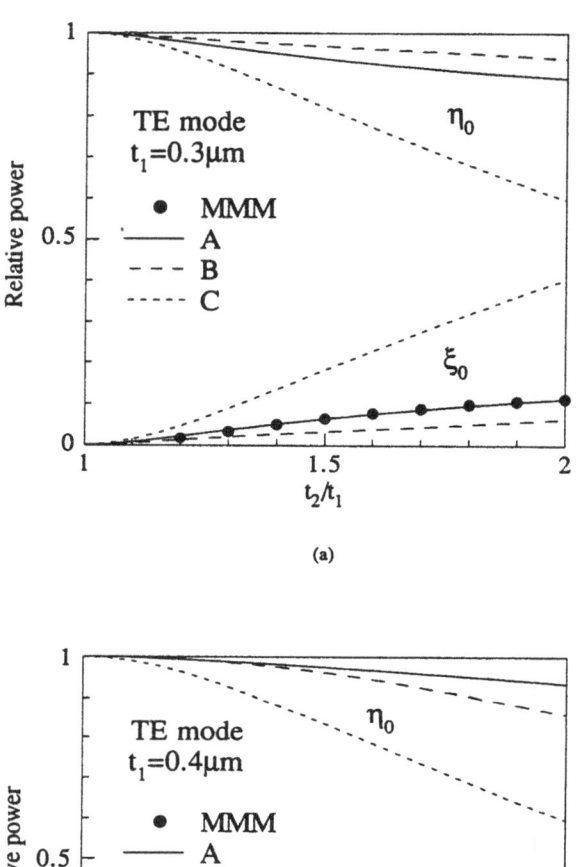

Figure 7.4: Scattering characteristics of a step for the TE mode incidence. (a) $t_1 = 0.3$ μm. (b) $t_1 = 0.4$ μm.

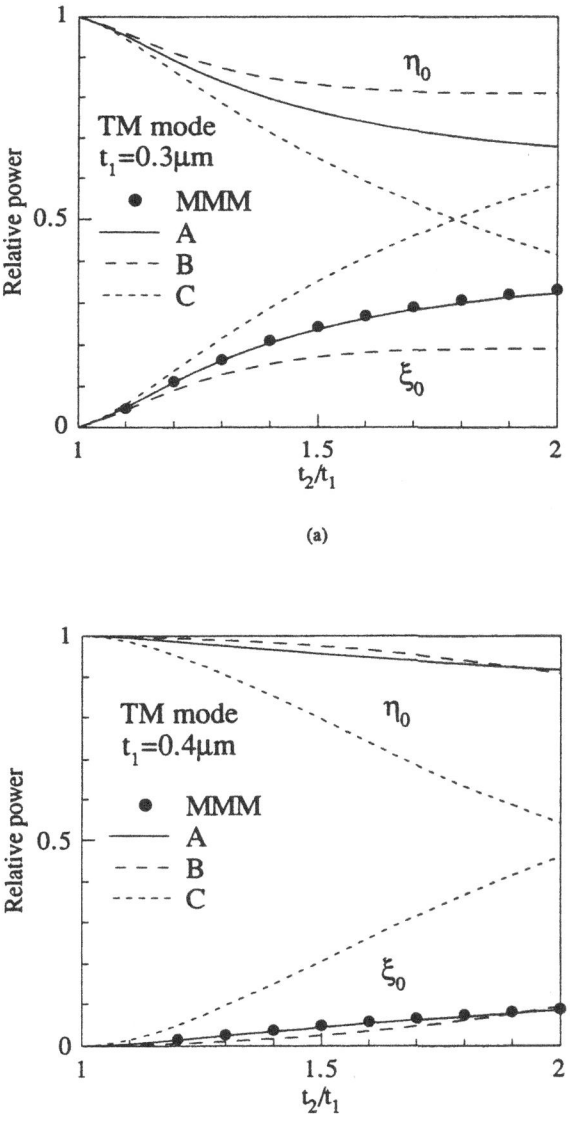

(a)

(b)

Figure 7.5: Scattering characteristics of a step for the TM mode incidence. (a) $t_1 = 0.3\ \mu$m. (b) $t_1 = 0.4\ \mu$m.

(4) Odd radiation modes ($r = 2$):

$$
h_2(y, \rho) =
\begin{cases}
\pm \left[\sin \kappa_f \dfrac{t}{2} \cos \rho \left(|y| - \dfrac{t}{2} \right) \right. \\
\qquad \left. + \dfrac{p_f \kappa_f}{p_s \rho} \cos \kappa_f \dfrac{t}{2} \sin \rho \left(|y| - \dfrac{t}{2} \right) \right] & \text{for } |y| \geq t/2 \\
\sin \kappa_f y & \text{for } |y| \leq t/2.
\end{cases}
\tag{7.58}
$$

where the double signs + and − are for the regions $y \geq t/2$ and $y \leq -t/2$, respectively, and the characteristic equation for the propagation constant β_m of the guided modes is given by

$$
\tan \kappa_{fm} t/2 =
\begin{cases}
p_s \alpha_{sm} / p_f \kappa_{fm} & \text{for even modes} \\
-p_f \kappa_{fm} / p_s \alpha_{sm} & \text{for odd modes.}
\end{cases}
\tag{7.59}
$$

The far-field radiation pattern in waveguide i ($i = 1, 2$; see Fig. 7.1) can be evaluated approximately by the method of stationary phase[21] as

$$
\begin{aligned}
I_i(\theta) = \frac{1}{2\beta_{01}} \left| \exp[-j\sigma_1(\rho)] \int_{-\infty}^{\infty} g_{1i}^*(y, \rho)\psi(x_i, y)\, dy \right. \\
\left. \pm j \exp[-j\sigma_2(\rho)] \int_{-\infty}^{\infty} g_{2i}^*(y, \rho)\psi(x_i, y)\, dy \right|^2
\end{aligned}
\tag{7.60}
$$

with

$$
\rho = \pm n_s k_0 \sin \theta
\tag{7.61}
$$

$$
\sigma_r(\rho) =
\begin{cases}
\tan^{-1} \left[\dfrac{p_f \kappa_f(\rho)}{p_s \rho} \tan \kappa_f(\rho) \dfrac{t}{2} \right] & \text{for } r = 1 \\
\tan^{-1} \left[\dfrac{p_s \rho}{p_f \kappa_f(\rho)} \tan \kappa_f(\rho) \dfrac{t}{2} \right] & \text{for } r = 2
\end{cases}
\tag{7.62}
$$

where the double signs + and − are for the regions $0 < \theta < \pi$ and $-\pi < \theta < 0$, respectively. The radiation pattern in Eq. (7.60) is normalized by the input power.

First, we consider a symmetric step as shown in Fig. 7.6, where $t_2 = \lambda/\pi$, $n_f = \sqrt{5}$, $n_s = 1$, and the fundamental TE mode incidence is assumed. Figures 7.7 and 7.8 show the scattering characteristics and radiation pattern for the step, respectively. Comparing the results of the FBEM with those of the variational method (VM),[22] it is confirmed that, using the FBEM, we may evaluate the scattering characteristics including the radiation pattern with good accuracy.

Next, we consider a waveguide bend as shown in Fig. 7.9, where $t = 6\lambda$,

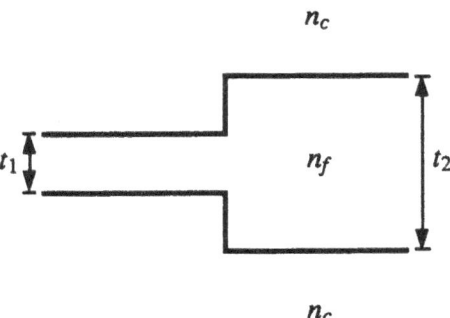

Figure 7.6: Step discontinuity.

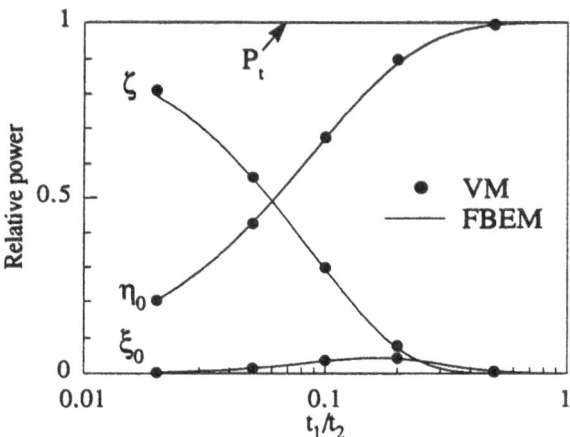

Figure 7.7: Scattering characteristics for a step.

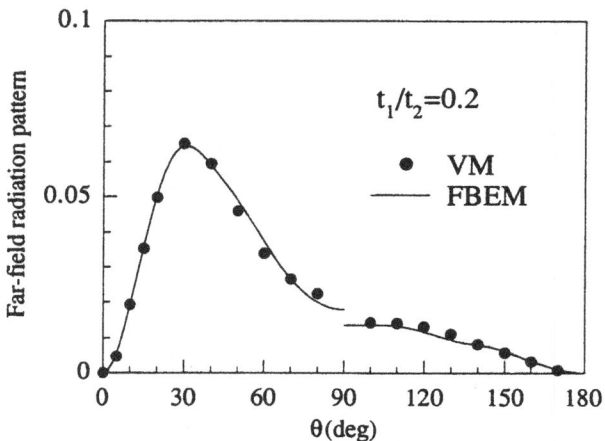

Figure 7.8: Far-field radiation pattern for a step.

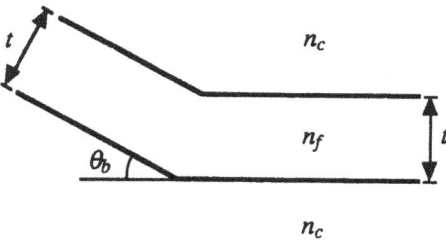

Figure 7.9: Waveguide bend.

$n_f = 1.502$, $n_s = 1.5$, and the fundamental TE mode incidence is assumed. Figures 7.10 shows the bending loss ($-10 \log \eta_0$ (dB)) as a function of the bending angle θ_b. The results obtained agree well with those of the beam propagation method (BPM).[23] The radiation patterns for $\theta_b = 2°$ and $4°$ are shown in Fig. 7.11.

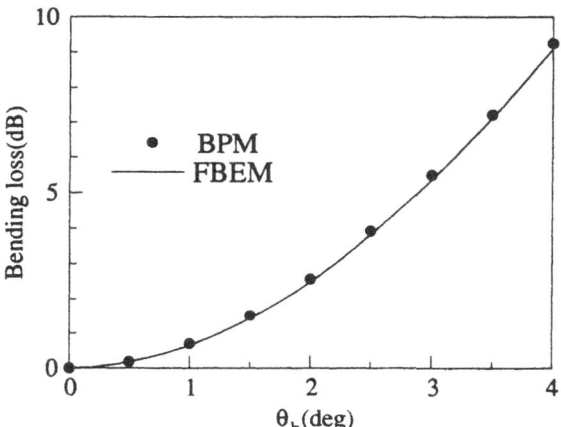

Figure 7.10: Bending loss.

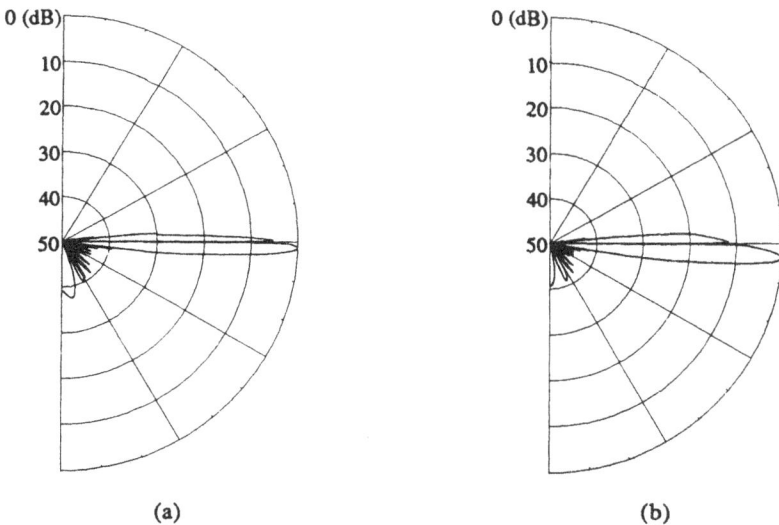

Figure 7.11: Far-field radiation pattern for a bend. (a) $\theta_b = 2°$. (b) $\theta_b = 4°$.

References

1) M. Suzuki and M. Koshiba, "Finite element analysis of discontinuity problems in a planar dielectric waveguide", *Radio Sci.*, Vol. 17, No. 1, pp. 85–91, Jan.-Feb. 1982.

2) M. Koshiba, K. Ooishi, T. Miki, and M. Suzuki, "Finite-element analysis of the discontinuities in a dielectric slab waveguide bounded by parallel plates", *Electron. Lett.*, Vol. 18, No. 1, pp. 33–34, Jan. 1982.

3) M. Koshiba, T. Miki, K. Ooishi, and M. Suzuki, "On finite-element solutions of the discontinuity problems in a bounded dielectric slab waveguide", *Trans. Inst. Electron. Commun. Eng. Japan*, Vol. E66, No. 4, pp. 250–251, April 1983.

4) M. Koshiba and M. Suzuki, "Boundary-element analysis of dielectric slab waveguide discontinuities", *Appl. Opt.*, Vol. 25, No. 6, pp. 828–829, March 1986.

5) M. Koshiba and K. Hirayama, "Application of finite element method to arbitrarily shaped discontinuities in a dielectric slab waveguide", *IEE Proc.*, Vol. 135, Pt. H, No. 1, pp. 8–12, Feb. 1988.

6) S.-J. Chung and C. H. Chen, "Partial variational principle for electromagnetic field problems: Theory and application", *IEEE Trans. Microwave Theory Tech.*, Vol. 36, No. 3, pp. 473–479, March 1988.

7) W. Kamminga and B. J. Hoenders, "Finite element solutions of waveguide-interconnection problems", *Opt. Commun.*, Vol. 67, No. 4, pp. 277–281, July 1988.

8) S.-J. Chung and C. H. Chen, "Analysis of irregularities in a planar dielectric waveguide", *IEEE Trans. Microwave Theory Tech.*, Vol. 36, No. 9, pp. 1352–1358, Sept. 1988.

9) S.-J. Chung and C. H. Chen, "A partial variational approach for arbitrary discontinuities in planar dielectric waveguides", *IEEE Trans. Microwave Theory Tech.*, Vol. 37, No. 1, pp. 208–214, Jan. 1989.

10) K. Hirayama and M. Koshiba, "Analysis of discontinuities in an open dielectric slab waveguide by combination of finite and boundary elements", *IEEE Trans. Microwave Theory Tech.*, Vol. 37, No. 4, pp. 761–768, April 1989.

11) B. M. Dillon, R. L. Ferrari, and T. P. Young, "Hybrid finite element/boundary element method for analysis of discontinuities in planar dielectric waveguides", *Electron. Lett.*, Vol. 26, No. 1, pp. 47–48, Jan. 1990.

12) K. Hirayama and M. Koshiba, "Numerical analysis of arbitrarily shaped discontinuities between planar dielectric waveguides with different thicknesses", *IEEE Trans. Microwave Theory Tech.*, Vol. 38, No. 3, pp. 260–264, March 1990.

13) W. Kamminga and B. J. Hoenders, "Experiences with finite element calculations of waveguide coupling problems", *Opt. Commun.*, Vol. 77, No. 5, 6, pp. 423-434, July 1990.

14) J. P. Webb, "Finite element methods for junctions of microwave and optical waveguides", *IEEE Trans. Magnet.*, Vol. 26, No. 5, pp. 1754–1758, Sept. 1990.

15) J. P. Webb, "Absorbing boundary conditions for the finite-element analysis of planar devices", *IEEE Trans. Microwave Theory Tech.*, Vol. 38, No. 9, pp. 1328–1332, Sept. 1990.

16) J. P. Webb and S. McFee, "The use of hierarchal triangles in finite-element analysis of microwave and optical devices", *IEEE Trans. Magnet.*, Vol. 27, No. 5, pp. 4040–4043, Sept. 1991.

17) K. Hirayama and M. Koshiba, "Analysis of discontinuities in an asymmetric dielectric slab waveguide by combination of finite and boundary elements", *IEEE Trans. Microwave Theory Tech.*, Vol. 40, No. 4, pp. 686–691, April 1992.

18) K. Hirayama and M. Koshiba, "A new low-loss structure of abrupt bends in dielectric waveguides", *IEEE/OSA Jour. Lightwave Technol.*, Vol. 10, No. 5, pp. 563–569, May 1992.

19) K. Hirayama and M. Koshiba, "Analysis of discontinuities in a grounded slab waveguide by combination of finite-element and boundary-element methods", *Int. Jour. Microwave and Millimeter-Wave Computer-Aided Engineering*, Vol. 2, No. 4, pp. 241–247, Oct. 1992.

20) V. Ramaswamy and P. G. Suchoski, Jr., "Power loss at a step discontinuity in an asymmetrical dielectric slab waveguide", *Jour. Opt. Soc. Am. A*, Vol. 1, No. 7, pp. 754–759, July 1984.

21) M. Born and E. Wolf, *Principles of Optics* (4th edn.), Pergamon Press, Oxford, 1970.

22) T. E. Rozzi, "Rigorous analysis of the step discontinuity in a planar dielectric waveguide", *IEEE Trans. Microwave Theory Tech.*, Vol. MTT-26, No. 10, pp. 738–746, Oct. 1978.

23) T. Shiina, K. Shiraishi, and S. Kawakami, "Waveguide-bend configuration with low-loss characteristics", *Opt. Lett.*, Vol. 11, No. 11, pp. 736–738, Nov. 1986.

Chapter 8

NONLINEAR OPTICAL WAVEGUIDES

8.1 Introduction

Optical waveguides are ideal for nonlinear interactions because they provide strong beam confinement over long propagation distances.[1] Progress in nonlinear guided-wave optics has concentrated on phenomena depending on the second-order and third-order nonlinear susceptibilities.

Many second-order guided-wave phenomena have been demonstrated, including second-harmonic generation (SHG), sum-frequency generation (SFG), difference-frequency generation (DFG), optical parametric amplification, and optical parametric oscillation. Recently there has been much research interest in guided-wave SHG devices to implement compact short-wavelength coherent light sources which would be useful tools in many applications, e.g., optical data storage, xerography, spectroscopy, and space and underwater communication.

On the other hand, third-order guided-wave phenomena have also been demonstrated, including intensity-dependent refractive-index phenomena, parametric mixing, degenerate four-wave mixing, and stimulated Raman and Brillouin scattering. In particular, applications of intensity-dependent refractive indices caused by the optical Kerr-effect nonlinearity are of great interest because of their unique features and their potential use in all-optical signal processing devices, such as bistability, switching, and upper and lower threshold devices and optical limiters.

In this chapter the FEM is described for the analysis of nonlinear guided-wave problems, placing emphasis on the SHG and intensity-dependent refractive-index phenomena.[2]−[22] Through a variety of numerical examples for Cerenkov-type SHG devices and Kerr-type nonlinear waveguides, we demonstrate that the FEM is useful for designing nonlinear guided-wave optical devices.

8.2   Method of Analysis for Second-Order Guided-Wave Phenomena

8.2.1   Basic equations

We consider an optical channel waveguide having a diagonal permittivity tensor and the second-order nonlinear optical tensor $[d]$ (see Eq. (1.17)). Implying the use of a scalar wave approximation (see Section 3.5) and considering a single linear eigenmode as an incident pump wave, an explicit form of the total optical field component may be written as

$$
\begin{aligned}
\Phi(x, y, z, t) = & \ \{\phi_1(x, y) \exp[j(\omega t - \beta z)] \\
& + \phi_2(x, y, z) \exp[j(2\omega t - 2\beta z)] \\
& + \text{c.c.}\}/2
\end{aligned}
\tag{8.1}
$$

with

$$
\Phi = \begin{cases} E_x & \text{for } E^x \text{ modes} \\ H_x & \text{for } E^y \text{ modes} \end{cases}
\tag{8.2}
$$

where ϕ_2 is the slowly varying amplitude of the second-harmonic (SH) field, and ω and β are, respectively, the angular frequency and the propagation constant of the fundamental wave. It should be noted that when we consider a single linear eigenmode as an incident pump wave, the fundamental field ϕ_1 is independent of z.

Noting that the E^x and E^y modes are well approximated by the TE^y ($E_y \equiv 0$, a leading function is E_x) and TM^y ($H_y \equiv 0$, a leading function is H_x) modes, respectively, from Maxwell's equations the following coupled nonlinear wave equations are derived:

$$
p_{x1} \frac{\partial^2 \phi_1}{\partial x^2} + p_{y1} \frac{\partial^2 \phi_1}{\partial y^2} - p_{z1} \beta^2 \phi_1 + q_1 k_0^2 \phi_1 = 0
\tag{8.3}
$$

$$
-j4\beta p_{z2} \frac{\partial \phi_2}{\partial z} + p_{x2} \frac{\partial^2 \phi_2}{\partial x^2} + p_{y2} \frac{\partial^2 \phi_2}{\partial y^2} - 4 p_{z2} \beta^2 \phi_2 + 4 q_2 k_0^2 \phi_2 = \psi_{\text{NL}}
\tag{8.4}
$$

where the subscripts $j = 1$ and 2 designate the quantities for the fundamental and SH waves, respectively, and the coefficients p_{xj}, p_{yj}, p_{zj}, q_j, and the source field ψ_{NL} are written as follows:

(1) E^x modes:

$$
p_{xj} = n_{xj}^2 / n_{zj}^2
\tag{8.5a}
$$

$$
p_{yj} = 1
\tag{8.5b}
$$

$$
p_{zj} = 1
\tag{8.5c}
$$

$$
q_j = n_{xj}^2
\tag{8.5d}
$$

$$\psi_{\mathrm{NL}} = -\frac{2k_0^2}{\varepsilon_0}(i_x \cdot P_{\mathrm{NL}}) . \tag{8.6}$$

(2) E^y modes:

$$p_{xj} = 1/n_{yj}^2 \tag{8.7a}$$

$$p_{yj} = 1/n_{zj}^2 \tag{8.7b}$$

$$p_{zj} = 1/n_{yj}^2 \tag{8.7c}$$

$$q_j = 1 \tag{8.7d}$$

$$\psi_{\mathrm{NL}} = \frac{k_0}{\varepsilon_0 Z_0}\left[\frac{2\beta(i_y \cdot P_{\mathrm{NL}})}{n_{y2}^2} - j\frac{1}{n_{z2}^2}\frac{\partial(i_z \cdot P_{\mathrm{NL}})}{\partial y}\right] . \tag{8.8}$$

Here i_x, i_y, and i_z are the unit vectors in the x, y, and z directions, respectively, and the second-order nonlinear polarization P_{NL} is given by

$$P_{\mathrm{NL}} = \begin{bmatrix} P_x \\ P_y \\ P_z \end{bmatrix} = \varepsilon_0 \begin{bmatrix} d_{11} & d_{12} & d_{13} & d_{14} & d_{15} & d_{16} \\ d_{21} & d_{22} & d_{23} & d_{24} & d_{25} & d_{26} \\ d_{31} & d_{32} & d_{33} & d_{34} & d_{35} & d_{36} \end{bmatrix} \begin{bmatrix} e_{x1}^2 \\ e_{y1}^2 \\ e_{z1}^2 \\ 2e_{y1}e_{z1} \\ 2e_{z1}e_{x1} \\ 2e_{x1}e_{y1} \end{bmatrix} \tag{8.9}$$

with

$$E_1 = \begin{bmatrix} E_{x1} \\ E_{y1} \\ E_{z1} \end{bmatrix} = \frac{1}{2}\begin{bmatrix} e_{x1}(x,y)\exp[j(\omega t - \beta z)] + \mathrm{c.c.} \\ e_{y1}(x,y)\exp[j(\omega t - \beta z)] + \mathrm{c.c.} \\ e_{z1}(x,y)\exp[j(\omega t - \beta z)] + \mathrm{c.c.} \end{bmatrix} . \tag{8.10}$$

In the derivation of Eq. (8.4), the slowly varying envelope approximation (SVEA), $|\partial^2\phi_2/\partial z^2| \ll |4\beta p_{z2}\partial\phi_2/\partial z|$, has been employed, and the pump depletion has been neglected. The amplitudes of the fundamental electric fields, e_{x1}, e_{y1}, e_{z1}, are expressed approximately as

$$e_{x1} = \phi_1 \tag{8.11a}$$

$$e_{y1} = 0 \tag{8.11b}$$

$$e_{z1} = 0 \tag{8.11c}$$

for the E^x modes, and

$$e_{x1} = 0 \tag{8.12a}$$

$$e_{y1} = -\frac{Z_0\beta}{n_{y1}^2 k_0}\phi_1 \tag{8.12b}$$

$$e_{z1} = j\frac{Z_0}{n_{z1}^2 k_0}\frac{\partial\phi_1}{\partial y} \tag{8.12c}$$

for the E^y modes.

In the case of a planar waveguide ($\partial/\partial x = 0$), Eqs. (8.3) and (8.4) give exact expressions for two-dimensional nonlinear guided-wave problems.

8.2.2   Finite element approach

Dividing the waveguide cross section into a number of quadratic triangular elements (see Fig. 1.3(b)), we expand the fields ϕ_1 and ϕ_2 in each element as

$$\phi_1(x, y) = \{N(x, y)\}^{\mathrm{T}}\{\phi_1\}_e \tag{8.13}$$

$$\phi_2(x, y, z) = \{N(x, y)\}^{\mathrm{T}}\{\phi_2(z)\}_e \,. \tag{8.14}$$

Applying the standard FEM to Eqs. (8.3) and (8.4) yields

$$[K_1]\{\phi_1\} - \beta^2[M_1]\{\phi_1\} = \{0\} \tag{8.15}$$

$$-j4\beta[M_2]\frac{d\{\phi_2\}}{dz} + ([K_2] - 4\beta^2[M_2])\{\phi_2\} = \{\psi_{\mathrm{NL}}\} \tag{8.16}$$

with

$$[K_1] = \sum_e \iint_e [q_1 k_0^2 \{N\}\{N\}^{\mathrm{T}} - p_{x1}\{N_x\}\{N_x\}^{\mathrm{T}}$$
$$- p_{y1}\{N_y\}\{N_y\}^{\mathrm{T}}]\, dx\, dy \tag{8.17}$$

$$[M_1] = \sum_e \iint_e p_{z1}\{N\}\{N\}^{\mathrm{T}}\, dx\, dy \tag{8.18}$$

$$[K_2] = \sum_e \iint_e [4q_2 k_0^2 \{N\}\{N\}^{\mathrm{T}} - p_{x2}\{N_x\}\{N_x\}^{\mathrm{T}}$$
$$- p_{y2}\{N_y\}\{N_y\}^{\mathrm{T}}]\, dx\, dy \tag{8.19}$$

$$[M_2] = \sum_e \iint_e p_{z2}\{N\}\{N\}^{\mathrm{T}}\, dx\, dy \tag{8.20}$$

where the components of the $\{\psi_{\mathrm{NL}}\}$ vector are the values of ψ_{NL} at nodes in the waveguide cross section, and the homogeneous Dirichlet or Neumann boundary conditions are imposed on edges of the computational window (xy plane).

Equation (8.15) is a standard eigenvalue problem whose eigenvalue and eigenvector correspond to β^2 and $\{\phi_1\}$, respectively. In order to solve Eq. (8.16), the following split-step procedure (SSP) will be introduced:

(1) Step 1 (propagation effect):

$$-j4\beta[M_2]\frac{d\{\phi_2^{(1)}\}}{dz} + ([K_2] - 4\beta^2[M_2])\{\phi_2^{(1)}\} = \{0\} \,. \tag{8.21}$$

(2) Step 2 (nonlinear effect):

$$\{\phi_2^{(2)}\} = \{\phi_2^{(1)}\} + \int j\frac{1}{4\beta p_{z2}}\{\psi_{\mathrm{NL}}\}\,dz \qquad (8.22)$$

where the superscripts 1 and 2 designate the fields for steps 1 and 2, respectively.

Applying the finite difference method (FDM) to Eq. (8.21) yields the following matrix equation within a short interval $i\Delta z \le z < (i+1)\Delta z$ along the propagation direction $(i = 0, 1, 2, \cdots)$:

$$-j4\beta[M_2](\{\phi_2^{(1)}\}_{i+1} - \{\phi_2^{(1)}\}_i)/\Delta z$$
$$+ ([K_2] - 4\beta^2[M_2])[\theta\{\phi_2^{(1)}\}_{i+1} + (1-\theta)\{\phi_2^{(1)}\}_i] = \{0\} \quad (8.23)$$

which is reduced to

$$[L(\theta)]\{\phi_2^{(1)}\}_{i+1} = [L(\theta - 1)]\{\phi_2^{(1)}\}_i \qquad (8.24)$$

with

$$[L(\theta)] = -j4\beta[M_2] + \theta\Delta z([K_2] - 4\beta^2[M_2]) \qquad (8.25)$$

where θ is an artificial parameter $(0 \le \theta \le 1)$; $\theta = 0$, $1/2$, 1 for forward difference, Crank-Nicolson (CN), and backward difference schemes, respectively. The first scheme is conditionally stable, whereas the other two are unconditionally stable, indicating that the choice of the value is not crucial, provided that the value is at least within $1/2 \le \theta \le 1$. Hence in the following calculations we shall adopt the CN scheme.

The difference version of Eq. (8.22) is

$$\{\phi_2^{(2)}\}_{i+1} = \{\phi_2^{(1)}\}_{i+1} + j\frac{\Delta z}{4\beta p_{z2}}\{\psi_{\mathrm{NL}}\}. \qquad (8.26)$$

Substituting Eq. (8.24) into Eq. (8.26), we finally obtain

$$\{\phi_2\}_{i+1} = [L(\theta)]^{-1}[L(\theta - 1)]\{\phi_2\}_i + j\frac{\Delta z}{4\beta p_{z2}}\{\psi_{\mathrm{NL}}\} \qquad (8.27)$$

where $\{\phi_2\}_i \equiv \{\phi_2^{(1)}\}_i$ and $\{\phi_2\}_{i+1} \equiv \{\phi_2^{(2)}\}_{i+1}$ are the initial and final SH fields within the interval $i\Delta z \le z < (i+1)\Delta z$.

The eigenvalue and eigenvector of Eq. (8.15) are utilized to evaluate the SH field $\{\phi_2\}$ through successive calculations of Eq. (8.27). With $\{\phi_2\}$ being computed, the total SH power can then be evaluated by

$$P_2 = \begin{cases} \dfrac{\beta}{2Z_0k_0}\{\phi_2\}^\dagger[M_2]\{\phi_2\} & \text{for } E^x \text{ modes} \\[3mm] \dfrac{Z_0\beta}{2k_0}\{\phi_2\}^\dagger[M_2]\{\phi_2\} & \text{for } E^y \text{ modes} \end{cases} \qquad (8.28)$$

where † denotes a complex conjugate transpose.

It should be noted here that for stationary analysis we obtain the following matrix equation by setting $d/dz \equiv 0$ in Eq. (8.16):

$$([K_2] - 4\beta^2[M_2])\{\phi_2\} = \{\psi_{\text{NL}}\} \tag{8.29}$$

where the SH field $\{\phi_2\}$ is independent of z.

8.3 Method of Analysis for Third-Order Guided-Wave Phenomena

8.3.1 Basic equations

We consider a Kerr-law self-focusing nonlinear optical channel waveguide whose nonlinear relative permittivity tensor is given by

$$[\varepsilon_r] = \begin{bmatrix} n_x^2 & 0 & 0 \\ 0 & n_y^2 & 0 \\ 0 & 0 & n_z^2 \end{bmatrix} \tag{8.30}$$

with

$$
\begin{aligned}
n_x^2 &= n^2 + a(|E_x|^2 + |E_y|^2 + |E_z|^2) & (8.31\text{a}) \\
n_y^2 &= n^2 + a(|E_x|^2 + |E_y|^2 + |E_z|^2) & (8.31\text{b}) \\
n_z^2 &= n^2 + a(|E_x|^2 + |E_y|^2 + |E_z|^2) & (8.31\text{c})
\end{aligned}
$$

where n is the linear refractive index and a is the nonlinear coefficient.

Under the scalar wave approximation and the SVEA, from Maxwell's equations the following equation is derived:

$$-j2\beta \frac{\partial \phi}{\partial z} + p_x \frac{\partial^2 \phi}{\partial x^2} + p_y \frac{\partial^2 \phi}{\partial y^2} - p_z \beta^2 \phi + q k_0^2 \phi = 0 \tag{8.32}$$

where β/k_0 is the reference (background) index, and the slowly varying amplitude ϕ and the coefficients p_x, p_y, p_z, q are written as follows:

(1) E^x modes:

$$E_x = \{\phi(x, y, z) \exp[j(\omega t - \beta z)] + \text{c.c.}\}/2 \tag{8.33}$$

$$
\begin{aligned}
p_x &= n_x^2/n_z^2 & (8.34\text{a}) \\
p_y &= 1 & (8.34\text{b}) \\
p_z &= 1 & (8.34\text{c}) \\
q &= n_x^2 . & (8.34\text{d})
\end{aligned}
$$

(2) E^y modes:

$$H_x = \{\phi(x, y, z) \exp[j(\omega t - \beta z)] + \text{c.c.}\}/2 \qquad (8.35)$$

$$p_x = 1/n_y^2 \qquad (8.36\text{a})$$
$$p_y = 1/n_z^2 \qquad (8.36\text{b})$$
$$p_z = 1/n_y^2 \qquad (8.36\text{c})$$
$$q = 1. \qquad (8.36\text{d})$$

In the case of a planar waveguide ($\partial/\partial x = 0$), Eq. (8.32) gives an exact expression for two-dimensional nonlinear guided-wave problems.

8.3.2 Finite element approach

Dividing the waveguide cross section into a number of quadratic triangular elements (see Fig. 1.3(b)), we expand the fields ϕ in each element as

$$\phi(x, y, z) = \{N(x, y)\}^{\mathrm{T}}\{\phi(z)\}_e . \qquad (8.37)$$

Applying the standard FEM to Eqs. (8.32) yields

$$- j2\beta[M]\frac{d\{\phi\}}{dz} + ([K] - \beta^2[M])\{\phi\} = \{0\} \qquad (8.38)$$

with

$$
\begin{aligned}
[K] &= \sum_e \iint_e [qk_0^2\{N\}\{N\}^{\mathrm{T}} - p_x\{N_x\}\{N_x\}^{\mathrm{T}} \\
&\quad - p_y\{N_y\}\{N_y\}^{\mathrm{T}}]\, dx\, dy \qquad (8.39) \\
[M] &= \sum_e \iint_e p_z\{N\}\{N\}^{\mathrm{T}}\, dx\, dy \qquad (8.40)
\end{aligned}
$$

where the homogeneous Dirichlet or Neumann boundary conditions are imposed on edges of the computational window (xy plane).

Applying the FDM to Eq. (8.38) yields the following matrix equation within a short interval $i\Delta z \le z < (i+1)\Delta z$ along the propagation direction ($i = 0, 1, 2, \cdots$):

$$
\begin{aligned}
&-j2\beta[M]_i(\{\phi\}_{i+1} - \{\phi\}_i)/\Delta z \\
&+ ([K]_i - \beta^2[M]_i)[\theta\{\phi\}_{i+1} + (1 - \theta)\{\phi\}_i] = \{0\} \qquad (8.41)
\end{aligned}
$$

which is reduced to

$$[L(\theta)]_i\{\phi\}_{i+1} = [L(\theta - 1)]_i\{\phi\}_i \qquad (8.42)$$

with

$$[L(\theta)]_i = -j2\beta[M]_i + \theta\Delta z([K]_i - \beta^2[M]_i) \qquad (8.43)$$

where θ is introduced as an artifice to control computational stability ($0 \leq \theta \leq 1$; $\theta = 1/2$ for the CN scheme, which is adopted in our numerical implementation).

Step-by-step calculations of Eq. (8.42) for $i = 0, 1, 2, \cdots$ yield the evolutional variation of the field profile fired at the entrance ($z = 0$) of the waveguide. Once E_x or H_x is identified, other field components are obtainable through Eq. (3.76) or Eq. (3.79). Among them E_x, E_y, and E_z are subsequently used to obtain the nonlinear response of material (see Eq. (8.31)).

It should be noted that when $n_x^2 = n_z^2$, for the E^x modes the following SSP is extremely useful for enhancing computational efficiencies:

(1) Step 1 (propagation effect):

$$- j2\beta[M]\frac{d\{\phi^{(1)}\}}{dz} + [K_0]\{\phi^{(1)}\} = \{0\} \qquad (8.44)$$

with

$$[K_0] = \sum_e \iint_e [-\{N_x\}\{N_x\}^T - \{N_y\}\{N_y\}^T]\, dx\, dy . \qquad (8.45)$$

(2) Step 2 (phase-rotating effect):

$$\{\phi^{(2)}\} = \{\phi^{(1)}\} \exp\left(-j\int_z k\, dz\right) \qquad (8.46)$$

with

$$k = \frac{qk_0^2 - \beta^2}{2\beta} \qquad (8.47)$$

where the superscripts 1 and 2 designate the fields for steps 1 and 2, respectively.

Applying the FDM to Eq. (8.44) yields an equation similar to Eq. (8.42):

$$[L(\theta)]\{\phi^{(1)}\}_{i+1} = [L(\theta - 1)]\{\phi^{(1)}\}_i \qquad (8.48)$$

but the matrix $[L(\theta)]$ is replaced by

$$[L(\theta)] = -j2\beta[M] + \theta\Delta z[K_0] . \qquad (8.49)$$

The difference version of Eq. (8.46) is

$$\{\phi^{(2)}\}_{i+1} = \{\phi^{(1)}\}_{i+1} \exp(-jk_i\Delta z) . \qquad (8.50)$$

Substituting Eq. (8.48) into Eq. (8.50), we finally obtain

$$\{\phi\}_{i+1} = [L(\theta)]^{-1}[L(\theta - 1)]\{\phi\}_i \exp(-jk_i\Delta z) \qquad (8.51)$$

where $\{\phi\}_i \equiv \{\phi^{(1)}\}_i$ and $\{\phi\}_{i+1} \equiv \{\phi^{(2)}\}_{i+1}$ are the initial and final fields within the interval $i\Delta z \leq z < (i+1)\Delta z$. This SSP results in a significant reduction of computational effort because the matrix $[L(\theta)]^{-1}[L(\theta - 1)]$ in Eq. (8.51) is linear and independent of z.

It should be noted here that for stationary analysis we obtain the following matrix eigenvalue problem by setting $d/dz \equiv 0$ in Eq. (8.38):

$$[K]\{\phi\} - \beta^2[M]\{\phi\} = \{0\} \qquad (8.52)$$

where the field $\{\phi\}$ is independent of z.

Self-consistent solutions of Eq. (8.52) which can be obtained by using the iterative scheme[4] give intensity-dependent propagation constants of nonlinear optical waveguides.

8.4 Cerenkov-Type Second-Harmonic Generation

We consider the Cerenkov-type phase matching in a proton-exchanged (PE) step-index channel waveguide schematically sketched in Fig. 8.1(a). Below we concentrate our attention on a reduced geometry in Fig. 8.1(b), which corresponds to the side view of actual three-dimensional structures, implicitly including a specific channel width W through an application of the effective refractive-index approximation. In accordance with the experimental optimization,[23] we adopt $W = 2.0$ μm as a guide width. Wavelength dependence of the ordinary (n_o) and extraordinary (n_e) refractive indices of the LiNbO$_3$ (LN) substrate is given by[24]

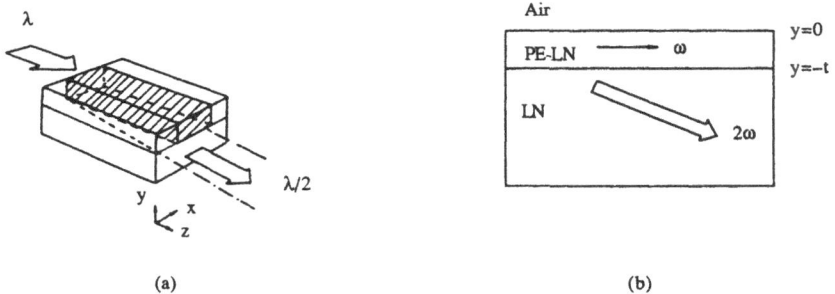

(a)   (b)

Figure 8.1: Schematics of Cerenkov SHG in proton-exchanged channel waveguide. (a) Three-dimensional view. (b) Side view.

$$n_o^2 = 4.9048 - \frac{0.11768}{0.04750 - \lambda^2} - 0.027169\lambda^2 \qquad (8.53a)$$

$$n_e^2 = 4.5820 - \frac{0.099169}{0.044432 - \lambda^2} - 0.021950\lambda^2 \qquad (8.53b)$$

where λ is the wavelength in μm. For $c//x$, $n_x = n_e$ and $n_y = n_z = n_o$; for $c//y$, $n_x = n_z = n_o$ and $n_y = n_e$; for $c//z$, $n_x = n_y = n_o$ and $n_z = n_e$. Here, c denotes the crystalline axis (optical axis). Typical values of maximum changes of the ordinary ($\Delta n_{o,\max}$) and extraordinary ($\Delta n_{e,\max}$) refractive indices of the PE-LN waveguides are listed in Table 8.1.[23),25)]

Table 8.1: Maximum index changes of the PE-LN waveguide.

λ (μm)	$\Delta n_{o,\max}$	$\Delta n_{e,\max}$
0.42	−0.064	0.190
0.53	−0.054	0.163
0.84	−0.043	0.130
1.06	−0.040	0.121

The explicit forms for the second-order nonlinear optical tensor of the LN crystal are written as[26),27)]

$$[d] = \begin{bmatrix} d_{33} & d_{31} & d_{31} & 0 & 0 & 0 \\ 0 & d_{22} & -d_{22} & 0 & 0 & d_{15} \\ 0 & 0 & 0 & -d_{22} & d_{15} & 0 \end{bmatrix} \quad \text{for } c//x \quad (8.54a)$$

$$[d] = \begin{bmatrix} 0 & 0 & 0 & 0 & -d_{22} & d_{15} \\ d_{31} & d_{33} & d_{31} & 0 & 0 & 0 \\ -d_{22} & 0 & d_{22} & d_{15} & 0 & 0 \end{bmatrix} \quad \text{for } c//y \quad (8.54b)$$

$$[d] = \begin{bmatrix} 0 & 0 & 0 & 0 & d_{15} & -d_{22} \\ -d_{22} & d_{22} & 0 & d_{15} & 0 & 0 \\ d_{31} & d_{31} & d_{33} & 0 & 0 & 0 \end{bmatrix} \quad \text{for } c//z \quad (8.54c)$$

with

$$d_{15} = d_{31} = -5.9 \times 10^{-12} \text{ m/V}$$
$$d_{22} = -4.0 \times 10^{-12} \text{ m/V}$$
$$d_{33} = -34 \times 10^{-12} \text{ m/V}.$$

We include nonlinearity in the channel as well as in the substrate, assume that the magnitude of d_{ij} does not change by the PE process, and ignore wavelength dependence because no explicit datum is available for the estimation.

Numerical convergence of simulated results is checked by changing mesh parameters in an adaptive fashion. Throughout our simulations we employ hundreds of quadratic line elements (one-dimensional analysis) and choose $\Delta z = 0.1\lambda$ as a longitudinally marching step. With these discretizations we have ensured satisfactory convergence of the computed results. In the numerical examples shown below we assume that the principal electric-field vectors of both interacting waves are equally aligned to the single crystalline axis, c, of the LN crystal with standard orientation, i.e., $c//x$ for the TE-like mode, and $c//y$ for the TM-like mode.

The evolution of SH power pumped by a TM_0 fundamental wave is shown in Fig. 8.2, where as representative fundamental wavelengths we consider $\lambda = 0.84$ μm in Fig. 8.2(a) and $\lambda = 1.06$ μm in Fig. 8.2(b), which imply, respectively, a GaAs semiconductor laser diode and a Nd:YAG laser, P_1 is the pump-wave power, and n_{y2} is the linear refractive index of the LN substrate for the y axis at the harmonic wavelength.

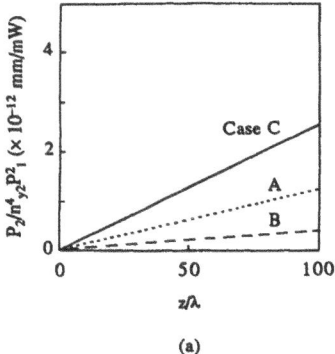 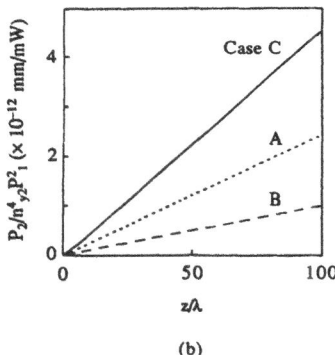

(a)   (b)

Figure 8.2: Total SH power versus interaction length (TM mode operation). (a) $\lambda = 0.84$ μm and $t = 0.35$ μm. (b) $\lambda = 1.06$ μm and $t = 0.525$ μm.

In Fig. 8.2 we consider three cases with respect to the nonlinearity of the film $(-t < y \leq 0)$: (case A) a linear film, i.e., $[d]_{\text{film}} = 0$; (case B) a nonlinear film with the same magnitude and sign of nonlinear susceptibility in the substrate that ranges over $y \leq -t$, i.e., $[d]_{\text{film}} = [d]_{\text{substrate}}$; and (case C) a domain-inverted film with the opposite sign of nonlinear susceptibility in the substrate, i.e., $[d]_{\text{film}} = -[d]_{\text{substrate}}$.[28] All the cases considered here are expected to occur, depending on the actual fabrication processes. For instance, case A may be seen in the situation in which degradation of the ideal crystalline structure in the guiding channel cannot be ignored. On the other hand, case C can be realized by adequately poling a certain kind of ferroelectric crystal, such as c-cut LN. Calculations have been implemented at the respective optimal thicknesses of the film, $t = 0.35$ μm in Fig. 8.2(a) and $t = 0.525$ μm in Fig. 8.2(b), the values of which were determined through the approximate analytical approach.[29] It is evident

from the figures that the SH power grows linearly with increasing interaction length and that the greatest SH power is obtainable in case C, the
value of which attains 6.5 and 4.5 times that of case B for $\lambda = 0.84$ μm
and $\lambda = 1.06$ μm, respectively. This significant enhancement of the SHG
can be attributed mathematically to the increasing overlap quantified by
the integral of the product of the nonlinear polarization wave (source) and
the SH wave (radiation) and physically to the increasing possibility of the
in-phase interference between the two. The present results show that in
Cerenkov doubling the phase-matching requirement is important not only
along the z axis but along the y axis, because the SH wave propagates
in the yz plane with its wave front tilted with the Cerenkov angle. As
an illustrative example of the SH field evolution, we show in Fig. 8.3 the
electric-field distribution that corresponds to the results of Fig. 8.2(b). The
results shown in Fig. 8.3 are consistent with those obtained in Fig. 8.2(b).

8.5   Kerr-Type Nonlinear Waveguides

We consider a planar optical waveguide with a Kerr-law self-focusing
nonlinear substrate,[30] where linear refractive indices of the cover ($y \geq t/2$),
the film ($|y| \leq t/2$), and the substrate ($y \leq -t/2$) are, respectively, $n_c =$
1.569, $n_f = 1.571$, and $n_s = 1.570$; nonlinear coefficient of the substrate
$a = 10^{-10}$ m^2/V^2; film thickness $t = 10$ μm; wavelength $\lambda = 0.633$ μm;
launching power $P = 2.92$ mW/mm, and the explicit form of the nonlinear
relative permittivities is

$$n_x^2 = n_s^2 + a|E_x|^2 \tag{8.55a}$$

$$n_y^2 = n_s^2 + a(|E_y|^2 + |E_z|^2) \tag{8.55b}$$

$$n_z^2 = n_s^2 + a(|E_y|^2 + |E_z|^2) . \tag{8.55c}$$

As a reference index β/k_0, we adopt the effective index of the corresponding linear fundamental mode. The waveguide cross section, ranging over
$|y| \leq 50$ μm, is subdivided into a number of quadratic line elements with
100 elements and 201 nodes (one-dimensional analysis); the difference step
for the propagation direction is chosen as $\Delta z = \lambda$, and the CN method
is adopted as a difference scheme because of its computational stability.
With these discretizations we have ensured satisfactory convergence of the
computed results.

Figure 8.4 shows the evolutional variation of the nonlinear TE wave
down the guide, where the corresponding linear TE$_0$ mode is fired at the
entrance ($z = 0$). Figure 8.4(a) shows that solitons are emitted from the
linear film into the lossless nonlinear substrate as the wave propagates down
the guide. This phenomenon manifests the symmetry-breaking instabilities
and bifurcations inherent in nonlinear systems far from equilibrium. A

slight asymmetry is purposely induced in the structure, $n_s - n_c = 10^{-3}$, which assists the breaking of symmetry of the field. In contrast to this, no soliton is emitted within the propagation distance under investigation in Fig. 8.4(b) because of the dissipation of the lossy substrate. A number of minute ripples emerging as the wave propagates are due to reflections of the radiative fraction from the truncated boundaries located on $|y| = 50 \ \mu$m.

Figure 8.5 shows the evolutional variation of the nonlinear TM wave along a bent waveguide, where the corresponding linear TM_0 mode is fired at the entrance.

These results indicate that soliton emission is more readily activated by adiabatically bending the waveguide axis so that its center of curvature lies in the linear cladding.

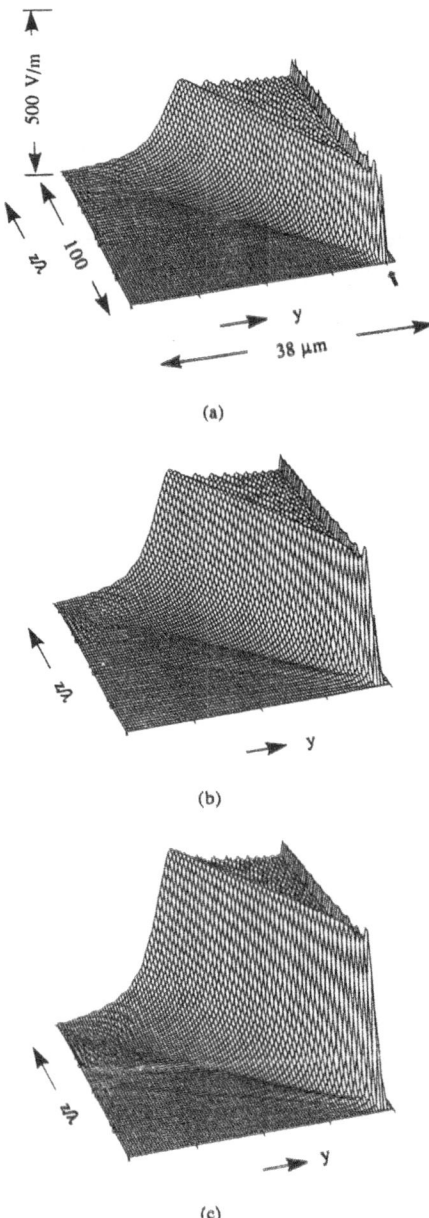

Figure 8.3: Evolution of SH field strength, pumped by the TM_0 mode with $P_1 = 10^4$ mW/mm ($\lambda = 1.06$ μm, $t = 0.525$ μm). (a) Case B. (b) Case A. (c) Case C.

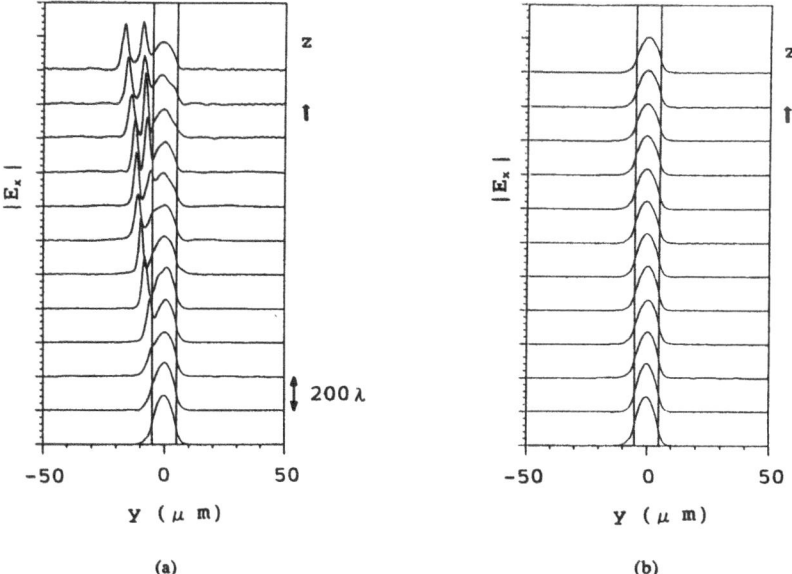

Figure 8.4: Evolutional variation of a nonlinear TE wave along a planar waveguide. (a) Transparent substrate ($\text{Im}(n_s)=0$). (b) Dissipative substrate ($\text{Im}(n_s)=-10^{-3}$).

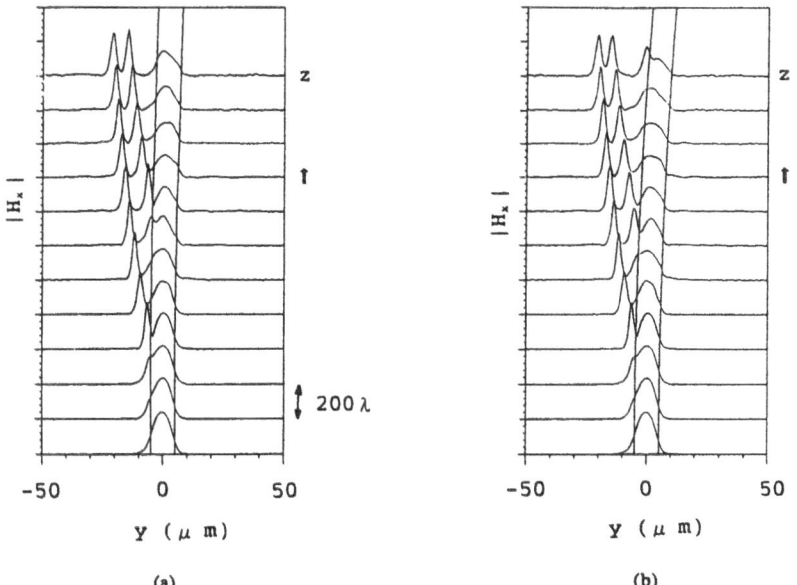

Figure 8.5: Evolutional variation of a nonlinear TM wave along a bent waveguide with radius of curvature R. (a) $R = 500$ mm. (b) $R = 200$ mm.

References

1) G. I. Stegeman and R. H. Stolen, "Waveguides and fibers for nonlinear optics", *Jour. Opt. Soc. Am. B*, Vol. 6, No. 4, pp. 652–662, April 1989.

2) K. Hayata, M. Koshiba, and M. Suzuki, "Finite-element solution of arbitrarily nonlinear, graded-index slab waveguides", *Electron. Lett.*, Vol. 23, No. 8, pp. 429–431, April 1987.

3) K. Hayata, M. Nagai, and M. Koshiba, "Finite-element theory of nonlinear TM-polarised guided waves", *Electron. Lett.*, Vol. 23, No. 24, pp. 1305–1307, Nov. 1987.

4) K. Hayata, M. Nagai, and M. Koshiba, "Finite-element formalism for nonlinear slab-guided waves", *IEEE Trans. Microwave Theory Tech.*, Vol. 36, No. 7, pp. 1207–1215, July 1988.

5) K. Hayata and M. Koshiba, "Self-focusing instability and chaotic behavior of nonlinear optical waves guided by dielectric slab structures", *Opt. Lett.*, Vol. 13, No. 11, pp. 1041–1043, Nov. 1988.

6) K. Hayata and M. Koshiba "Full vectorial analysis of nonlinear-optical waveguides", *Jour. Opt. Soc. Am. B*, Vol. 5, No. 12, pp. 2494–2501, Dec. 1988.

7) B. M. A. Rahman and J. B. Davies, "Finite element solution of nonlinear bistable optical waveguides", *Int. Jour. Optoelectronics*, Vol. 4, No. 2, pp. 153–161, 1989.

8) K. Hayata and M. Koshiba, "Numerical simulation of guided-wave SHG light sources utilising Cerenkov radiation scheme", *Electron. Lett.*, Vol. 25, No. 6, pp. 376–377, March 1989.

9) K. Hayata, A. Misawa, and M. Koshiba, "Nonlinear beam propagation in tapered waveguides", *Electron. Lett.*, Vol. 25, No. 10, pp. 661–662, May 1989.

10) K. Okamoto and E. A. J. Marcatili, "Chromatic dispersion characteristics of fibers with optical Kerr-effect nonlinearity", *IEEE/OSA Jour. Lightwave technol.*, Vol. 7, No. 12, pp. 1988–1994, Dec. 1989.

11) K. Hayata, A. Misawa, and M. Koshiba, "Nonstationary simulation of nonlinearly coupled TE-TM waves propagating down dielectric slab structures by the step-by-step finite-element method", *Opt. Lett.*, Vol. 15, No. 1, pp. 24–26, Jan. 1990.

12) A. D. Boardman, T. Twardowski, and E. M. Wright, "The effect of diffusion on surface-guided nonlinear TM waves: A finite element approach", *Opt. Commun.*, Vol. 74, No. 5, pp. 347–352, Jan. 1990.

13) B. M. A. Rahman, J. R. Souza, and J. B. Davies, "Numerical analysis of nonlinear bistable optical waveguides", *IEEE Photonics Technol. Lett.*, Vol. 2, No. 4, pp. 265–267, April 1990.

14) K. Hayata, A. Misawa, and M. Koshiba, "Chaotic propagation in the elliptically-polarized light in nonlinear guided-wave structures", *Trans. Inst. Electron. Inform. Commun. Eng.*, Vol. E73, No. 6, pp. 855–860, June 1990.

15) K. Hayata, A. Misawa, and M. Koshiba, "Spatial polarization instabilities due to transverse effects in nonlinear guided-wave systems", *Jour. Opt. Soc. Am. B*, Vol. 7, No. 7, pp. 1268–1280, July 1990.

16) K. Hayata and M. Koshiba, "Cerenkov third-harmonic generation in optical glass fibre", *Electron. Lett.*, Vol. 26, No. 19, pp. 1566–1567, Sept. 1990.

17) K. Hayata, A. Misawa, and M. Koshiba, "Split-step finite-element method applied to nonlinear integrated optics", *Jour. Opt. Soc. Am. B*, Vol. 7, No. 9, pp. 1772–1784, Sept. 1990.

18) K. Hayata and M. Koshiba, "Numerical study of guided-wave sum-frequency generation through second-order nonlinear parametric processes", *Jour. Opt. Soc. Am. B*, Vol. 8, No. 2, pp. 449–458, Feb. 1991.

19) R. D. Ettinger, F. A. Fernandez, B. M. A. Rahman, and J. B. Davies, "Vector finite element solution of saturable nonlinear strip-loaded optical waveguides", *IEEE Photonics Technol. Lett.*, Vol. 3, No. 2, pp. 147–149, Feb. 1991.

20) K. Hayata and M. Koshiba, "Mutual guiding assistance between eigenmodes of nonlinearly coupled TE-TM waves", *Trans. Inst. Electron. Inform. Commun. Eng.*, Vol. E74, No. 9, pp. 2890–2897, Sept. 1991.

21) K. Hayata and M. Koshiba, "Three-dimensional simulation of guided-wave second-harmonic generation in the form of coherent Cerenkov radiation", *Opt. Lett.*, Vol. 16, No. 20, pp. 1563–1565, Oct. 1991.

22) D. Mihalache and D. Mazilu, "Propagation phenomena of nonlinear guided waves in graded-index planar waveguides", *IEE Proc.*, Vol. 138, Pt. J, No. 6, pp. 365–372, Dec. 1991.

23) T. Taniuchi and K. Yamamoto, "Second harmonic generation in proton exchanged guides", *SPIE Proc.*, Vol. 864, pp. 36–41, 1987.

24) D. S. Smith, H. D. Riccius, and R. P. Edwin, "Refractive indices of lithium niobate", *Opt. Commun.*, Vol. 17, No. 3, pp. 332–335, June 1976.

25) M. J. Li, M. De Michell, Q. He, and D. B. Ostrowsky, "Cerenkov configuration second harmonic generation in proton-exchanged lithium niobate guides", *IEEE Jour. Quantum Electron.*, Vol. 26, No. 8, pp. 1384–1393, Aug. 1990.

26) A. Yariv and P. Yeh, *Optical Waves in Crystals*, Wiley, New York, 1984.

27) D. A. Kleinman, "Nonlinear dielectric polarization in optical media", *Phys. Rev.*, Vol. 126, No. 6, pp. 1977–1979, June 1962.

28) K. Hayata, T. Yanagawa, and M. Koshiba, "Enhancement of the guided-wave second-harmonic generation in the form of Cerenkov radiation", *Appl. Phys. Lett.*, Vol. 56, No. 3, pp. 206–208, Jan. 1990.

29) K. Hayata, T. Sugawara, and M. Koshiba, "Modal analysis of the second-harmonic electromagnetic field generated by the Cerenkov effect in optical waveguides", *IEEE Jour. Quantum Electron.*, Vol. 26, No. 1, pp. 123–134, Jan. 1990.

30) L. Leine, C. Wächter, U. Langbein, and F. Lederer, "Evolution of nonlinear guided optical fields down a dielectric film with a nonlinear cladding", *Jour. Opt. Soc. Am. B*, Vol. 5, No. 2, pp. 547–558, Feb. 1988.

Chapter 9

OPTICAL SOLITONS

9.1 Introduction

In the negative group velocity dispersion (GVD) region of optical fibers, solitons can be generated by balancing self-phase modulation (SPM) with negative GVD. A soliton is one of the key concepts that will support future optical communications with a capability for large-capacity, long-distance transmission systems. Solitons in optical fibers were first proposed by Hasegawa and Tappert, and they demonstrated the stability of the solitons by numerical calculations.[1] In 1980, Mollenauer *et al.* observed solitons in optical fibers.[2]

It is known that the soliton pulse propagation in optical fibers is described by the perturbed nonlinear Schrödinger equation (PNLSE). Without any perturbation, this equation is reduced to the canonical nonlinear Schrödinger equation (NLSE) and can be solved by means of the inverse scattering method (ISM),[3] but with nonzero additional perturbations such as dissipation (amplification), soliton self-frequency shift (SSFS),[4],[5] higher-order dispersion, and nonlinear dispersion, the equation cannot be solved exactly.

More recently, the effects of those perturbations on the soliton propagation have actually been observed in experiments with the generation of ultrashort pulses and the use of low-loss fibers, indicating that the effects can no longer be ignored, and there have been many numerical investigations of soliton propagation in connection with the transmission capacity of optical fiber communication systems.[6]−[13]

In this chapter the FEM is described for the analysis of soliton pulse propagation dynamics in an optical fiber.[14] First, to confirm the validity of this approach we analyze unperturbed problems and show that the results obtained are in good agreement with those derived from other approaches. In addition, we carry out investigations on the interaction between adjacent soliton pulses as well as on perturbations such as dissipation and SSFS. The results obtained show that the FEM is useful for predicting soliton dynamics in realistic optical fibers.

9.2   Method of Analysis

9.2.1   Basic equations

From Maxwell's equations the following vectorial wave equation is derived:

$$\nabla^2 \boldsymbol{E} - \mu_0 \frac{\partial^2 \boldsymbol{D}}{\partial t^2} = 0 \, . \tag{9.1}$$

Assuming that the incident light is polarized along a principle axis (chosen to coincide with the x axis), the electric field for the fundamental fiber mode (HE_{11} or LP_{01} mode) is approximately given by

$$\boldsymbol{E} = \boldsymbol{i}_x \{\phi(r) E(z,t) \exp[j(\omega t - \beta z)] + \text{c.c.}\}/2 \tag{9.2}$$

where $\phi(r)$ is the transverse-field function (see Eq. (4.36)), $E(z,t)$ is the complex amplitude function, and $\boldsymbol{i}_x$ is the unit vector in the x direction.

The intensity-dependent refractive index of a fiber is given by[7]

$$n(\lambda, |\boldsymbol{E}|^2) = n(\lambda) + n_{\mathrm{NL}} |\boldsymbol{E}|^2 \tag{9.3}$$

with

$$n_{\mathrm{NL}} = \frac{3}{8n} \chi_{xxxx} \tag{9.4}$$

where λ is the wavelength of free space, n_{NL} is the nonlinear-index (Kerr) coefficient which for silica is 3.2×10^{-20} m^2/W (1.22×10^{-22} m^2/V^2), and χ_{xxxx} is the third-order nonlinear optical susceptibility.

Considering Eqs. (9.2) and (9.3) into Eq. (9.1) and using the slowly varying envelope approximation (SVEA) for the propagation direction, we obtain the following PNLSE describing the propagation of high-intensity optical pulse envelopes in an optical fiber with negative GVD and self-focusing nonlinearity:

$$-j\frac{\partial q}{\partial \xi} + \frac{1}{2}\frac{\partial^2 q}{\partial s^2} + |q|^2 q - j\Gamma q - R\frac{\partial |q|^2}{\partial s} q = 0 \tag{9.5}$$

with

$$q = \sqrt{\frac{4\pi^2 c \tau^2 \tilde{n}_{\mathrm{NL}}}{\lambda^3 |\sigma|}} \, E \tag{9.6}$$

$$\xi = \frac{\lambda^2 |\sigma|}{2\pi c\tau^2} z = \frac{\pi}{2}\frac{z}{z_s} \tag{9.7}$$

$$s = \frac{1}{\tau}\left(t - \frac{z}{v_g}\right) \tag{9.8}$$

$$\Gamma = \frac{2}{\pi} z_s \frac{\alpha}{20000 \log_{10} e} \tag{9.9}$$

$$R = \frac{\tau_R}{\tau} \tag{9.10}$$

$$\tilde{n}_{NL} = n_{NL} \int_0^{2\pi}\int_0^\infty \phi^4 r\, dr\, d\theta \Big/ \int_0^{2\pi}\int_0^\infty \phi^2 r\, dr\, d\theta \tag{9.11}$$

$$\frac{1}{v_g} = \frac{\partial\beta}{\partial\omega} \tag{9.12}$$

$$\sigma = \frac{2\pi c}{\lambda^2}\frac{\partial^2\beta}{\partial\omega^2} \tag{9.13}$$

$$z_s = \frac{\pi^2 c\tau^2}{\lambda^2 |\sigma|} = 0.322 \frac{\pi^2 c\tau_{\text{FWHM}}^2}{\lambda^2 |\sigma|} \tag{9.14}$$

$$\tau_{\text{FWHM}} = 1.76\tau \tag{9.15}$$

where q represents a normalized complex amplitude of the pulse envelope, ξ is a normalized distance along the fiber, s is a Galilean time with the frame of reference moving with the linear group velocity of the waveguide, c is the light velocity (see Eq. (4.64)), v_g is the group velocity, $\sigma < 0$ is the negative GVD, α (dB/km) is the fiber loss coefficient, τ is the normalizing time, τ_{FWHM} is the FWHM of the pulse intensity, the parameter τ_R governs the retarded nonlinear response, and z_s corresponding to $\xi = \pi/2$ is called the soliton period. The response time τ_R is estimated to be about 5.9 fs for silica fibers.[15] In Eq. (9.5), for simplicity, the higher-order (linear and nonlinear) dispersions[7] are neglected.

When fiber loss and SSFS are ignored ($\Gamma = R = 0$), Eq. (9.5) is reduced to the canonical NLSE. The NLSE has a steady state and periodic solutions. The initial condition is

$$q(s) = A \operatorname{sech} s \tag{9.16}$$

which is well known as the lowest soliton where $1/2 < A < 3/2$.

Assuming the transverse-field function $\phi(r)$ takes the form

$$\phi(r) \propto \exp[-(r/W)^2] \tag{9.17}$$

the peak power of the pulse required to generate an exact $N = 1$ soliton pulse ($A = 1$) is given by

$$P_1 = 0.776 \frac{\lambda^3 |\sigma|}{\pi^2 c n_{NL} \tau_{\text{FWHM}}^2} A_{\text{eff}} \tag{9.18}$$

with

$$A_{\text{eff}} = \left(\int_0^{2\pi} \int_0^{\infty} \phi^2 r \, dr \, d\theta \right)^2 \Big/ \int_0^{2\pi} \int_0^{\infty} \phi^4 r \, dr \, d\theta = \pi W^2 \qquad (9.19)$$

where W is the mode field radius and the parameter A_{eff} is known as the effective core area.

9.2.2 Finite element approach

Dividing the s direction of Eq. (9.5), namely, the Galilean time domain, into a number of quadratic line elements (see Fig. 1.2(b)), we expand the normalized complex amplitude of the pulse envelope, q, in each element as

$$q(\xi, s) = \{N(s)\}^{\text{T}} \{q(\xi)\}_e . \qquad (9.20)$$

Applying the standard FEM to Eq. (9.5) yields

$$- j[M] \frac{d\{q\}}{d\xi} + [K]\{q\} = \{0\} \qquad (9.21)$$

with

$$[K] = \sum_e \int_e \left[-\frac{1}{2} \{N_s\}\{N_s\}^{\text{T}} + |q|^2 \{N\}\{N\}^{\text{T}} \right.$$
$$\left. - j\Gamma\{N\}\{N\}^{\text{T}} - R\frac{\partial |q|^2}{\partial s}\{N\}\{N\}^{\text{T}} \right] ds \qquad (9.22)$$

$$[M] = \sum_e \int_e \{N\}\{N\}^{\text{T}} ds \qquad (9.23)$$

where the periodic boundary conditions are imposed on edges of the computational window (s axis).

Applying the finite difference method (FDM) to Eq. (9.21) yields the following matrix equation within a short interval $i\Delta\xi \leq \xi < (i+1)\Delta\xi$ along the propagation direction ($i = 0, 1, 2, \cdots$):

$$- j[M](\{q\}_{i+1} - \{q\}_i)/\Delta\xi + [K]_i[\theta\{q\}_{i+1} + (1 - \theta)\{q\}_i] = \{0\} \qquad (9.24)$$

which is reduced to

$$[L(\theta)]_i\{q\}_{i+1} = [L(\theta - 1)]_i\{q\}_i \qquad (9.25)$$

with

$$[L(\theta)]_i = -j[M] + \theta\Delta\xi[K]_i \qquad (9.26)$$

where θ is introduced as an artifice to control computational stability ($0 \leq \theta \leq 1$; $\theta = 1/2$ for the Crank-Nicolson (CN) scheme, which is adopted in our numerical implementation).

It should be noted that the following split-step procedure (SSP) is extremely useful for enhancing computational efficiencies:

(1) Step 1 (dispersion effect):

$$-j[M]\frac{d\{q^{(1)}\}}{d\xi} + [K_0]\{q^{(1)}\} = \{0\} \qquad (9.27)$$

with

$$[K_0] = \sum_e \int_e \left[-\frac{1}{2}\{N_s\}\{N_s\}^{\mathrm{T}} \right] ds . \qquad (9.28)$$

(2) Step 2 (nonlinear effect):

$$\{q^{(2)}\} = \{q^{(1)}\} \exp\left(-j\int kd\xi\right) \qquad (9.29)$$

with

$$k = |q|^2 - j\Gamma - R\partial|q|^2/\partial s \qquad (9.30)$$

where the superscripts 1 and 2 designate the fields for steps 1 and 2, respectively.

Applying the FDM to Eq. (9.27) yields an equation similar to Eq. (9.25):

$$[L(\theta)]\{q^{(1)}\}_{i+1} = [L(\theta - 1)]\{q^{(1)}\}_i \qquad (9.31)$$

but the matrix $[L(\theta)]$ is replaced by

$$[L(\theta)] = -j[M] + \theta\Delta\xi[K_0] . \qquad (9.32)$$

The difference version of Eq. (9.29) is

$$\{q^{(2)}\}_{i+1} = \{q^{(1)}\}_{i+1} \exp(-jk_i\Delta\xi) . \qquad (9.33)$$

Substituting Eq. (9.31) into Eq. (9.33), we finally obtain

$$\{q\}_{i+1} = [L(\theta)]^{-1}[L(\theta - 1)]\{q\}_i \exp(-jk_i\Delta\xi) \qquad (9.34)$$

where $\{q\}_i \equiv \{q^{(1)}\}_i$ and $\{q\}_{i+1} \equiv \{q^{(2)}\}_{i+1}$ are the initial and final fields within the interval $i\Delta\xi \leq \xi < (i + 1)\Delta\xi$. This SSP results in a significant reduction of computational effort because the matrix $[L(\theta)]^{-1}[L(\theta - 1)]$ included in Eq. (9.34) is linear and independent of ξ.

9.3 Solitons in an Unperturbed Optical Fiber

As an initial condition on $\xi = 0$, consider a hyperbolic secant pulse with amplitude A (see Eq. (9.16)). In order to check the validity of the method, we first consider the unperturbed case ($\Gamma = R = 0$).

Figures 9.1(a), (b), and (c) show the results for $A = 1$ (fundamental soliton), 2 (second-order soliton), and 3 (third-order soliton), respectively, where N_E is the number of elements. For integral A, pure solitons propagate without any radiation. These results are in agreement with those predicted by the ISM.[3]

To examine the accuracy of the method, Table 9.1 shows the relative errors of the following conserved quantities[3] I_1, I_2 at $\xi = \pi$:

Table 9.1: Relative errors for the conserved quantities at $\xi = \pi$.

A	$|\Delta I_1|/I_1$ (%)	$|\Delta I_2|/I_2$ (%)
1	0.0000002	0.0004488
2	0.0000010	0.0000154
3	0.0000709	0.0011874

$$I_1 = \int_{-\infty}^{\infty} |q(\xi, s)|^2 \, ds \qquad (9.35)$$

$$I_2 = \int_{-\infty}^{\infty} [|\partial q(\xi, s)/\partial s|^2 - |q(\xi, s)|^4/2] \, ds . \qquad (9.36)$$

From Table 9.1 it is seen that the errors are negligible, indicating that satisfactorily accurate results are obtained.

We next apply the split-step FEM to an interaction between solitons. Mutual interference between adjacent pulses is worth studying to determine the maximum transmission rate (bit rate) in a practical optical communication system.

We consider the interaction between two pulses, which are fired at the entrance ($\xi = 0$) as

$$q(s) = \text{sech}(s - T/2) + \exp(j\theta_r)\text{sech}(s + T/2) \qquad (9.37)$$

where T and θ_r indicate, respectively, pulse separation and relative phase difference between adjacent pulses.

Figures 9.2 and 9.3 show the effect of the phase difference between two pulses. It is seen that two solitons coalesce periodically into a single pulse.

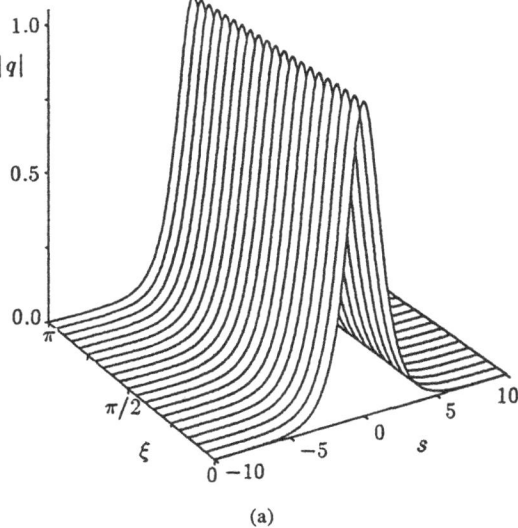

(a)

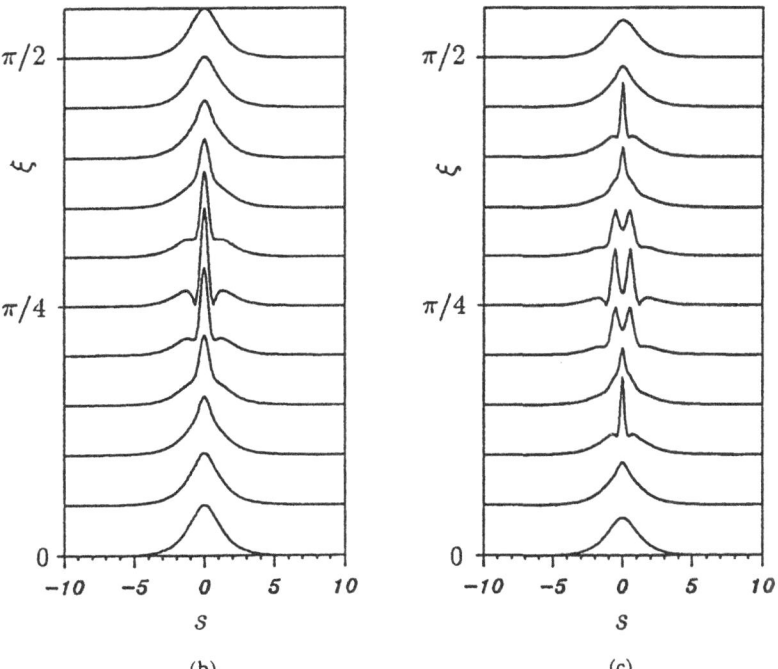

(b)   (c)

Figure 9.1: Soliton propagation for integral initial amplitude. (a) $A = 1$ ($N_E = 150$, $\Delta\xi = \pi/200$). (b) $A = 2$ ($N_E = 200$, $\Delta\xi = \pi/200$). (c) $A = 3$ ($N_E = 250$, $\Delta\xi = \pi/400$).

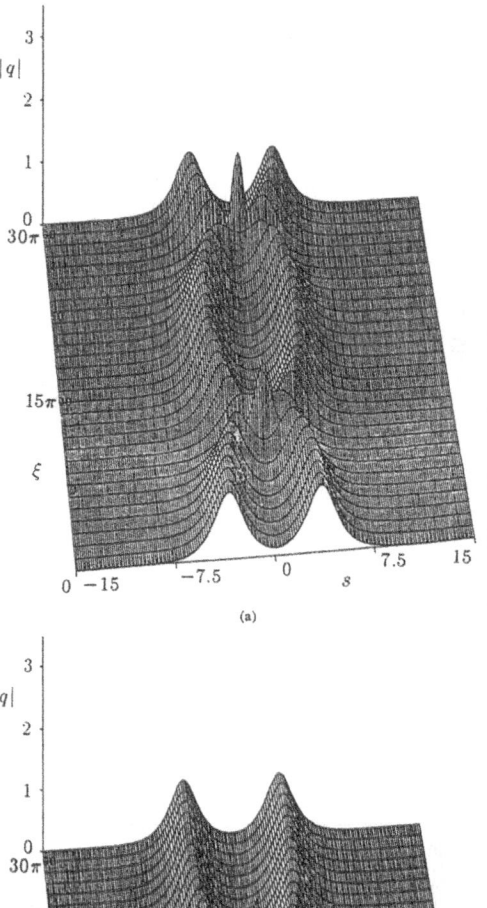

(a)

(b)

Figure 9.2: Soliton interaction for in-phase incidence ($N_E = 100$, $\Delta\xi = \pi/100$). (a) $T = 7$. (b) $T = 8$.

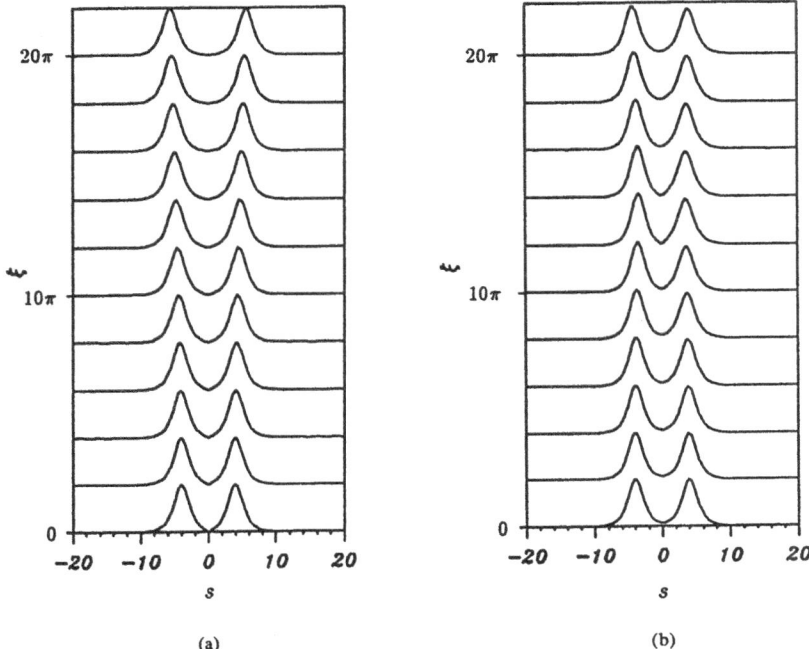

Figure 9.3: Soliton interaction for unequal-phase incidence ($N_E = 100$, $\Delta\xi = \pi/100$). (a) $\theta_r = \pi$ ($T = 8$). (b) $\theta_r = \pi/4$ ($T = 8$).

Moreover, we see that the wider separation between them causes the longer coalescent period, i.e., the longer spacing between repeaters. By contrast, for the out-of-phase ($\theta_r = \pi$) case, two pulses repulse each other, and their spacing increases as they propagate. In designing optical communication systems, it is necessary to preserve the spacing of adjacent pulses. When $\theta_r = \pi/4$, two pulses propagate nearly in parallel.[8)]

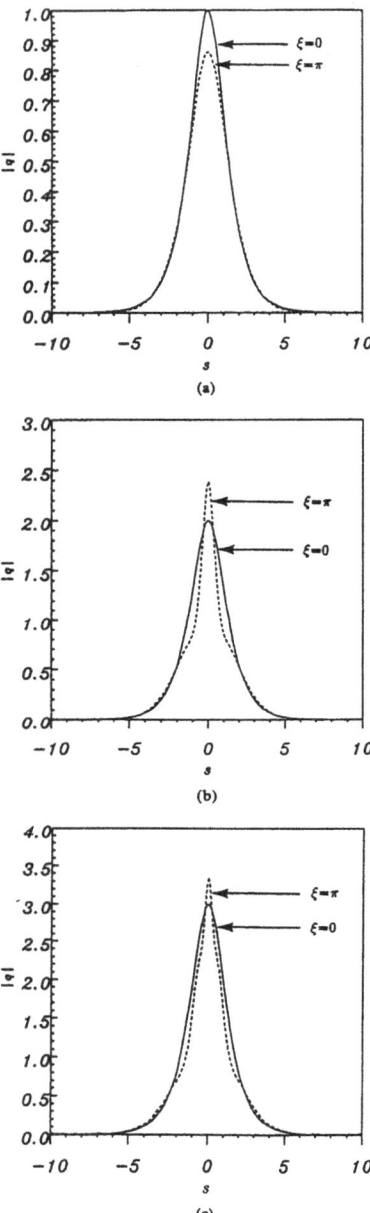

Figure 9.4: Effect of dissipation on soliton propagation ($\Gamma = 0.03$). (a) $A = 1$ ($N_E = 100$, $\Delta\xi = \pi/200$). (b) $A = 2$ ($N_E = 200$, $\Delta\xi = \pi/200$). (c) $A = 3$ ($N_E = 250$, $\Delta\xi = \pi/400$).

9.4 Solitons in a Perturbed Optical Fiber

It is interesting to see how pulse propagation is influenced by the material loss.

Figure 9.4 shows the effect of dissipation ($\Gamma = 0.03$) on the soliton propagation. The results agree well with those of the beam propagation method (BPM).[10]

Figure 9.5 shows the propagation to $\xi = 5\pi$ for $\Gamma = 0.035$. The pulse width broadens as a pulse propagates down the fiber because of weakened nonlinearity caused by decreasing amplitude. As a result, the amplitude at $\xi = 5\pi$ decreases 0.38 times relative to the incident one. In addition, the quantity given by Eq. (9.35) is 66.7 percent of the initial value, indicating an almost 30 percent energy loss within the domain $-15 \leq s \leq 15$.

Figure 9.6 shows the effect of dissipation on soliton interaction, where $\Gamma = 0.035$ and $T = 8$. These results are remarkably different from the unperturbed case (see Figs. 9.2 and 9.3).

Figures 9.7 and 9.8 correspond to cases including SSFS. Figure 9.7 shows the propagation of a single soliton. The group velocity v_g of the soliton decelerates due to the downshift of the center frequency caused by SSFS. For larger R, this effect is enhanced significantly.

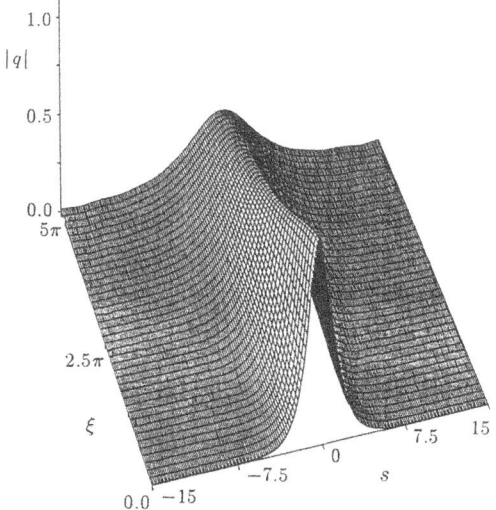

Figure 9.5: Evolution of the fundamental soliton in a lossy optical fiber ($N_E = 100$, $\Delta\xi = \pi/200$).

Next, to examine the influence of SSFS on soliton interaction, we analyze the case for $R = 0.01$ and $T = 5.288$. Figures 9.8(a), (b), and (c) correspond to phase differences, $\theta_r = 0$, -1, and 1 rad, respectively. In all cases two incident pulses collide and split again into two pulses.[9] From a comparison of these three cases, in-phase incidence is preferable with respect to the repeating space since it takes the longest distance to the collision. This is entirely opposite to the unperturbed case (see Figs. 9.2 and 9.3). Consequently, it is found that investigation on the suppression of interaction should be made under inclusion of perturbations.

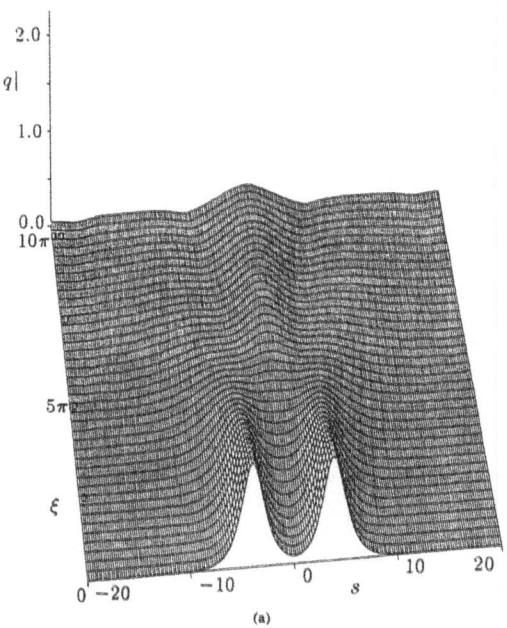

(a)

Fig. 9.6

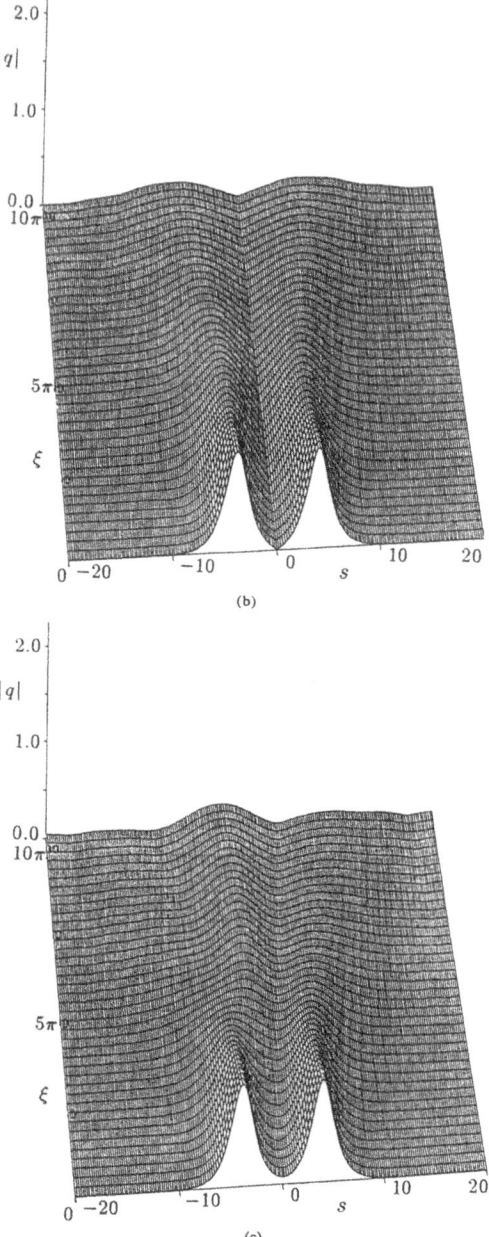

Figure 9.6: Effect of dissipation on soliton interaction ($N_E = 100$, $\Delta\xi = \pi/200$). (a) $\theta_r = 0$. (b) $\theta_r = \pi$. (c) $\theta_r = \pi/4$.

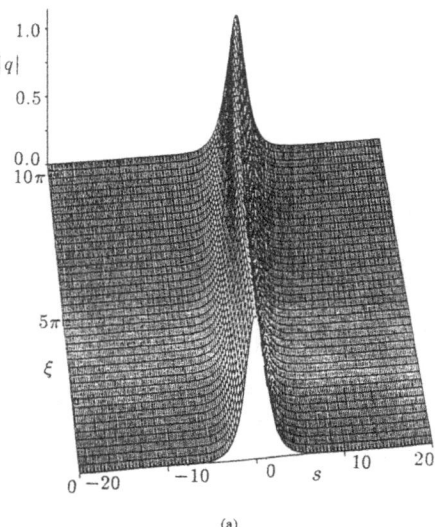

(a)

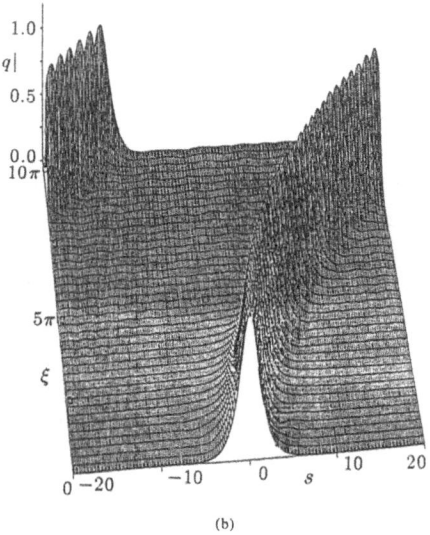

(b)

Figure 9.7: Effect of SSFS on soliton propagation ($N_E = 100$, $\Delta\xi = \pi/100$). (a) $R = 0.01$. (b) $R = 0.1$.

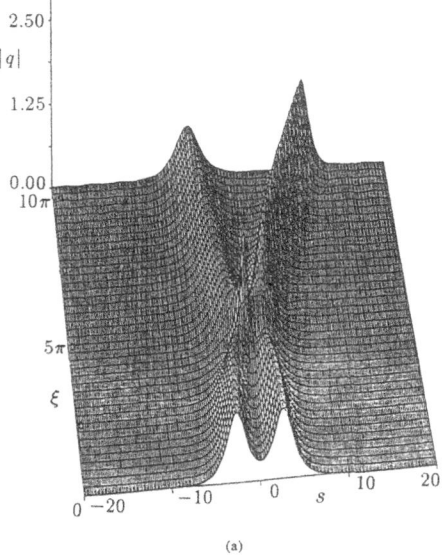

(a)

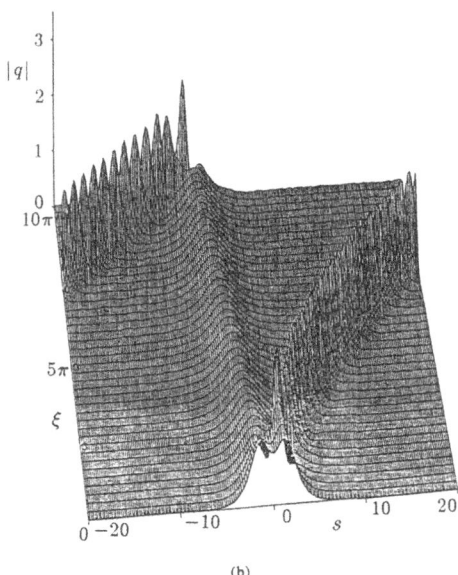

(b)

Fig. 9.8

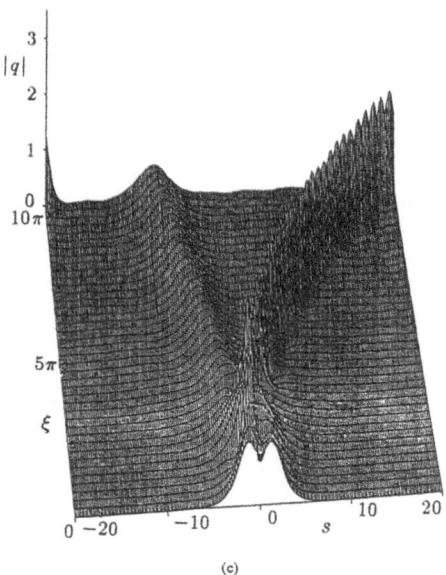

(c)

Figure 9.8: Effect of SSFS on soliton interaction ($N_E = 100$, $\Delta\xi = \pi/100$). (a) $\theta_r = 0$.
(b) $\theta_r = -1$ rad. (c) $\theta_r = +1$ rad.

References

1) A. Hasegawa and F. Tappert, "Transmission of stationary nonlinear optical pulses in dispersive dielectric fibers. I. Anomalous dispersion", *Appl. Phys. Lett.*, Vol. 23, No. 3, pp. 142–144, Aug. 1973.

2) L. F. Mollenauer, R. H. Stollen, and J. P. Gordon, "Experimental observation of picosecond pulse narrowing and solitons in optical fibers", *Phys. Rev. Lett.*, Vol. 45, No. 13, pp. 1095–1098, Sept. 1980.

3) J. Satsuma and N. Yajima, "Initial value problems of one-dimensional self-modulation of nonlinear waves in dispersive media", *Suppl. Prog. Theor. Phys.*, Vol. 55, pp. 284–306, 1974.

4) F. M. Mitschke and L. F. Mollenauer, "Discovery of the soliton self-frequency shift", *Opt. Lett.*, Vol. 11, No. 10, pp. 659–661, Oct. 1986.

5) J. P. Gordon, "Theory of soliton self-frequency shift", *Opt. Lett.*, Vol. 11, No. 10, pp. 662–664, Oct. 1986.

6) E. Shiojiri and Y. Fujii, "Transmission capability of an optical fiber communication system using index nonlinearity", *Appl. Opt.*, Vol. 24, No. 3, pp. 358–360, Feb. 1985.

7) Y. Kodama and A. Hasegawa, "Nonlinear pulse propagation in a monomode dielectric guide", *IEEE Jour. Quantum Electron.*, Vol. QE-23, No. 5, pp. 510–524, May 1987.

8) C. Desem and P. L. Chu, "Reducing soliton interaction in single-mode optical fibres", *IEE Proc.*, Vol. 134, Pt. J, No. 3, pp. 145–151, June 1987.

9) Y. Kodama and K. Nozaki, "Soliton interaction in optical fibers", *Opt. Lett.*, Vol. 12, No. 12, pp. 1038–1040, Dec. 1987.

10) Y. Zhao, "Soliton propagation in optical fibers with random perturbations", *Opt. Commun.*, Vol. 68, No. 1, pp. 21-24, Sept. 1988.

11) H. Kubota and M. Nakazawa, "Long-distance optical soliton transmission with lumped amplifiers", *IEEE Jour. Quantum Electron.*, Vol. 26, No. 4, pp. 692–700, April 1990.

12) K. Hayata, K. Saka, and M. Koshiba, "Path integral simulation of polarized solitons in optical fibers", *Jour. Appl. Phys.*, Vol. 68, No. 10, pp. 4971–4982, Nov. 1990.

13) M. Nakazawa, K. Suzuki, H. Kubota, E. Yamada, and Y. Kimura, "Dynamic optical soliton communication", *IEEE Jour. Quantum Electron.*, Vol. 26, No. 12, pp. 2095–2102, Dec. 1990.

14) M. Eguchi, K. Hayata, and M. Koshiba, "Effects of birefringence on the interaction between adjacent nonlinear pulses", *Opt. Lett.*, Vol. 16, No. 2, pp. 82–84, Jan. 1991.

15) H. A. Haus and M. Nakazawa, "Theory of the fiber Raman soliton laser", *Jour. Opt. Soc. Am. B*, Vol. 4, No. 5, pp. 652–660, May 1987.

Chapter 10

QUANTUM WELL STRUCTURES

10.1 Introduction

With the recent improvement in epitaxial crystal growth techniques such as molecular beam epitaxy (MBE) and metal organic chemical vapor deposition (MOCVD), it has become possible to fabricate high-quality layered semiconductor structures like quantum well (QW), quantum wire (QWI), quantum box (QB), superlattice (SL), and resonant tunneling (RT) structures.[1],[2] These structures have been attracting much attention in recent years because of their electronic and optoelectronic properties.

The quantum-mechanical properties of these structures are, in principle, obtained by solving the Schrödinger equation for the envelope function under the effective mass approximation. If the potential is constant or linear in a region, the Schrödinger equation can be solved analytically. However, if we consider an arbitrary potential profile, the Schrödinger equation must be solved numerically.

In this chapter the FEM is described for the analysis of QW, SL, and RT structures.[3]−[13] Mathematical formulations for investigating the eigenstates of QWs, the minibands of SLs, and the transmission probabilities of potential barriers are given, and the quantum-mechanical properties are evaluated for a variety of structures in the $GaAs/Al_xGa_{1-x}As$ system.

10.2 Method of Analysis

10.2.1 Basic equations

We consider a quantum well and/or potential barrier structure with an arbitrary potential $U(x)$ in the region $x_1 \leq x \leq x_2$. Under the effective mass approximation, the envelope function ϕ of an electron in the system is given by the one-dimensional time-independent Schrödinger equation

$$\frac{\hbar^2}{2} \frac{d}{dx} \left(\frac{1}{m_{\text{eff}}} \frac{d\phi}{dx} \right) - U\phi + E\phi = 0 \qquad (10.1)$$

with

$$\hbar = h/2\pi \qquad (10.2)$$

where m_{eff} and E are the effective mass and the energy of the electron, respectively, and $\hbar$ is the reduced Planck's constant. Planck's constant h is given by

$$h = 4.136 \times 10^{-15} \text{ eV·s} = 6.626 \times 10^{-34} \text{ J·s} . \qquad (10.3)$$

For the GaAs/Al$_x$Ga$_{1-x}$As system, the effective mass is given by

$$m_{\text{eff}} = (0.067 + 0.083x)m_0 \qquad (10.4)$$

with

$$m_0 = 9.110 \times 10^{-31} \text{ kg} \qquad (10.5)$$

where m_0 is the free electron mass. In the GaAs/AlGaAs system, the boundary condition has the well-known form that the envelope function and its derivative divided by the effective mass are continuous at the heterointerface.

The probability current density $S(x)$, at any x, of the quantum-mechanical system may be written as

$$S = \frac{1}{2jm_{\text{eff}}} \left(\phi \frac{d\phi^*}{dx} - \phi^* \frac{d\phi}{dx} \right) = \text{Re} \left(\frac{\phi\psi^*}{2jm_{\text{eff}}} \right) \qquad (10.6)$$

where $\psi \equiv d\phi/dx$.

The functional for Eq. (10.1) is given by

$$\begin{aligned} F &= \int_{x_1}^{x_2} \left(\frac{\hbar^2}{2m_{\text{eff}}} \frac{d\phi^*}{dx} \frac{d\phi}{dx} + U\phi^*\phi - E\phi^*\phi \right) dx \\ &\quad + \frac{\hbar^2}{2m_{\text{eff1}}} \phi_1^* \psi_1 - \frac{\hbar^2}{2m_{\text{eff2}}} \phi_2^* \psi_2 \end{aligned} \qquad (10.7)$$

where the subscripts 1 and 2 designate the values at $x = x_1$ and $x = x_2$, respectively.

10.2.2   Finite element approach

First, we consider a QW with an arbitrary potential $U(x)$ in the region $x_1 \leq x \leq x_2$, and with constant potentials U_1 and U_2 in the regions $x \leq x_1$ and $x \geq x_2$, respectively, as shown in Fig. 10.1.

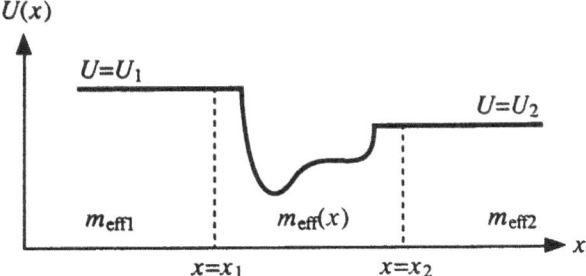

Figure 10.1: Potential profile of a quantum well.

Dividing the region $x_1 \leq x \leq x_2$ into a number of quadratic line elements (see Fig. 1.2(b)), we expand the envelope function ϕ in each element as

$$\phi = \{N\}^{\mathrm{T}}\{\phi\}_e \, . \tag{10.8}$$

Substituting Eq. (10.8) into Eq. (10.7), from the variational principle we obtain the following global matrix equation:

$$[K]\{\phi\} - E[M]\{\phi\} = \{\psi\} \tag{10.9}$$

with

$$[K] = \sum_e \int_e \left[\frac{\hbar^2}{2m_{\mathrm{eff}}}\{N_x\}\{N_x\}^{\mathrm{T}} + U\{N\}\{N\}^{\mathrm{T}} \right] dx \tag{10.10}$$

$$[M] = \sum_e \int_e \{N\}\{N\}^{\mathrm{T}} dx \tag{10.11}$$

$$\{\psi\} = \begin{bmatrix} -(\hbar^2/2m_{\mathrm{eff1}})\psi_1 \\ 0 \\ \vdots \\ (\hbar^2/2m_{\mathrm{eff2}})\psi_2 \end{bmatrix} \tag{10.12}$$

where the first and last components of the $\{\phi\}$ vector are the values of ϕ at nodes on $x = x_1$ and $x = x_2$, respectively.

The field ϕ in the uniform region $x \leq x_1$ or $x \geq x_2$ is given by

$$\phi \propto \begin{cases} \exp(\alpha_1 x) & \text{for } x \leq x_1 \\ \exp(-\alpha_2 x) & \text{for } x \geq x_2 \end{cases} \tag{10.13}$$

with

$$\alpha = \sqrt{2m_{\mathrm{eff}}(U - E)/\hbar^2} \tag{10.14}$$

where the subscripts 1 and 2 designate the solutions in the regions $x \leq x_1$ and $x \geq x_2$, respectively.

From Eq. (10.13) the values of ψ at $x = x_1$ and $x = x_2$ are given by

$$\psi_1 = \alpha_1 \phi_1 \tag{10.15}$$

$$\psi_2 = -\alpha_2 \phi_2 . \tag{10.16}$$

Substituting Eqs. (10.15) and (10.16) into Eq. (10.9), we obtain the final matrix equation:

$$[A]\{\phi\} = \{0\} \tag{10.17}$$

with

$$[A] = [\tilde{K}] - E[M] \tag{10.18}$$

$$[\tilde{K}] = [K] + \begin{bmatrix} \dfrac{\alpha_1 \hbar^2}{2m_{\text{eff1}}} & 0 & \cdots & 0 \\ 0 & 0 & \cdots & 0 \\ \vdots & \vdots & \ddots & \vdots \\ 0 & 0 & \cdots & \dfrac{\alpha_2 \hbar^2}{2m_{\text{eff2}}} \end{bmatrix} . \tag{10.19}$$

The condition that Eq. (10.17) has a nontrivial solution is given by

$$|A| = 0 \tag{10.20}$$

which is the proper equation for the eigenstates in a QW structure.

Setting artificial boundaries in the position far from the QW region and imposing the condition

$$\phi = 0 \tag{10.21}$$

or

$$d\phi/dx = 0 \tag{10.22}$$

on the artificial boundaries, $x = x_1$ and $x = x_2$, we obtain for Eq. (10.17)

$$[K]\{\phi\} - E[M]\{\phi\} = \{0\} . \tag{10.23}$$

Equation (10.23) is a standard eigenvalue problem whose eigenvalue and eigenvector correspond to E and $\{\phi\}$, respectively.

Next, we consider an SL with period d as shown in Fig. 10.2. From Floquet's theorem the periodic boundary conditions are given by

$$\phi_2 = \sigma \phi_1 \tag{10.24}$$

$$\frac{\hbar^2}{2m_{\text{eff2}}} \psi_2 = \sigma \frac{\hbar^2}{2m_{\text{eff1}}} \psi_1 \tag{10.25}$$

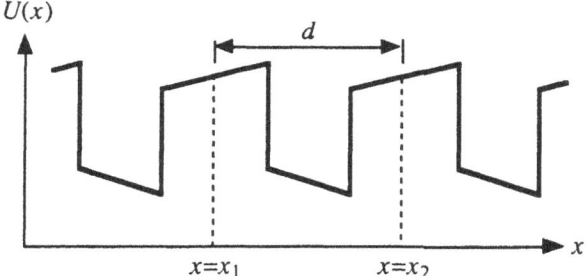

Figure 10.2: Potential profile of a superlattice.

with

$$\sigma = \exp(-j\beta d) \qquad (10.26)$$

where β is the wavenumber of the electron.

Considering the periodic boundary conditions, Eqs. (10.24) and (10.25), into Eq. (10.9), we obtain

$$[\tilde{K}]\{\phi\} - E[\tilde{M}]\{\phi\} = \{0\} \qquad (10.27)$$

with

$$[\tilde{K}] = [T]^{\dagger}[K][T] \qquad (10.28)$$
$$[\tilde{M}] = [T]^{\dagger}[M][T] \qquad (10.29)$$
$$[T] = \begin{bmatrix} 1 & 0 & \cdots & 0 \\ 0 & 1 & \cdots & 0 \\ \vdots & \vdots & \ddots & \vdots \\ 0 & 0 & \cdots & 1 \\ \sigma & 0 & \cdots & 0 \end{bmatrix} \qquad (10.30)$$

where $\dagger$ denotes a complex conjugate transpose.

By solving the eigenvalue equation (10.27), we can obtain the eigen-energy E as a function of the wavenumber β. The value of the envelope function ϕ at each node is also obtained at the same time.

Finally, we consider a potential barrier with an arbitrary potential $U(x)$ in the region $x_1 \le x \le x_2$, and with constant potentials U_1 and U_2 in the regions $x \le x_1$ and $x \ge x_2$, respectively, as shown in Fig. 10.3.

In the regions $x \le x_1$ and $x \ge x_2$, Eq. (10.1) can be solved analytically. Assuming that a plane wave of unit amplitude is incident from the left, the envelope function ϕ for $x \le x_1$ or $x \ge x_2$ is written as

$$\phi = \begin{cases} \exp(-j\beta_1 x) + b_1 \exp(j\beta_1 x) & \text{for } x \le x_1 \\ a_2 \exp(-j\beta_2 x) & \text{for } x \ge x_2 \end{cases} \qquad (10.31)$$

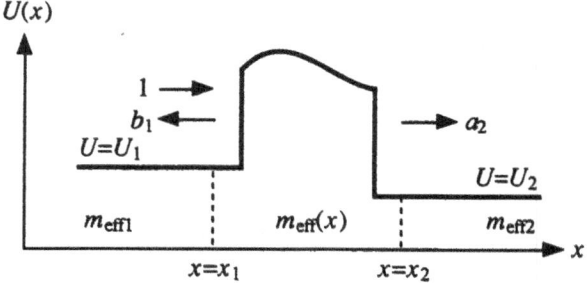

Figure 10.3: Potential profile of a potential barrier.

with

$$\beta = \sqrt{2m_{\text{eff}}(E-U)/\hbar^2} \tag{10.32}$$

where b_1 and a_2 are the amplitudes of reflected and transmitted waves, respectively. From Eq. (10.31) we may express the field ψ at $x = x_1$ and $x = x_2$ analytically as

$$\psi_1 = -j2\beta_1 \exp(-j\beta_1 x_1) + j\beta_1 \phi_1 \tag{10.33}$$
$$\psi_2 = -j\beta_2 \phi_2 . \tag{10.34}$$

Substituting Eqs. (10.33) and (10.34) into Eq. (10.9), we obtain the following matrix equation:

$$[A]\{\phi\} = \{\tilde{\psi}\} \tag{10.35}$$

with

$$[A] = [\tilde{K}] - E[M] \tag{10.36}$$

$$[\tilde{K}] = [K] + \begin{bmatrix} j\dfrac{\beta_1 \hbar^2}{2m_{\text{eff1}}} & 0 & \cdots & 0 \\ 0 & 0 & \cdots & 0 \\ \vdots & \vdots & \ddots & \vdots \\ 0 & 0 & \cdots & j\dfrac{\beta_2 \hbar^2}{2m_{\text{eff2}}} \end{bmatrix} \tag{10.37}$$

$$\{\tilde{\psi}\} = \begin{bmatrix} j\dfrac{\beta_1 \hbar^2}{m_{\text{eff1}}} \exp(-j\beta_1 x_1) \\ 0 \\ \vdots \\ 0 \end{bmatrix} . \tag{10.38}$$

The solutions of Eq. (10.35) allow the determination of the quantum mechanical reflection probability ξ and the quantum mechanical transmission probability η, as follows:

$$\xi = |\phi_1 - \exp(-j\beta_1 x_1)|^2 \qquad (10.39)$$

$$\eta = \frac{m_{\text{eff1}}\beta_2}{m_{\text{eff2}}\beta_1}|\phi_2|^2 . \qquad (10.40)$$

10.3 Eigenstates of a Quantum Well

In order to demonstrate the validity of the FEM, we first consider the eigenstates of a simple rectangular QW as shown in Fig. 10.4. Table 10.1 shows the eigenenergies of a rectangular QW with $U_w = 10E_1$ for the three levels $(n = 1, 2, 3)$, where $E_1 = \hbar^2\pi^2/2m_{\text{eff}}W^2$ corresponds to the eigenenergy of the ground state in a rectangular QW with well width W and an infinite well depth $(U_w = \infty)$, only the well region is divided into quadratic line elements, i.e., Eq. (10.20) is used, N_E is the number of elements, and the effective mass is assumed to be constant over the whole region. The results obtained agree well with the exact solutions.

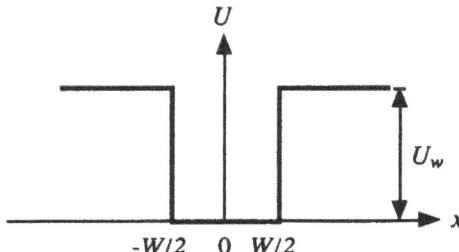

Figure 10.4: Rectangular quantum well.

Table 10.1: Eigenenergies of a rectangular QW (E/E_1).

n	Exact solution	FEM $N_E = 10$	$N_E = 20$
1	0.6901731	0.6901760	0.6901732
2	2.7239406	2.7241197	2.7239520
3	5.9495887	5.9513937	5.9497053

Next, let us consider the QW subject to an external electric field F normal to the QW layers. In this case, there exist no bound states in an exact sense because an electron tunnels out of the well. However, if this tunneling rate is small, the states become quasi-bound and we can consider discrete energy levels.[14]

Here, we consider only the finite region $x_1 \leq x \leq x_2$ and use Eq. (10.23), using the boundary conditions $\phi(x_1) = \phi(x_2) = 0$ at the edges of this region, which correspond to the infinite potential barriers outside this region. Therefore, when we calculate the eigenstates under this potential condition, the solutions would contain the eigenstates not only for the electron bound in the QW, but also for the electron bound near the end of this region, as shown in Fig. 10.5. Eigenenergies and envelope functions for $m = 1$ and 3 in Fig. 10.5 are due to the potential condition mentioned above. These eigenstates do not exist when we consider the actual QW without such infinite potential barriers. Hence, we should discard these eigenstates to obtain the correct quasi-bound states of the QW. The first ($n = 1$) quasi-bound state for this QW is obtained from the eigenstate for $m = 2$ in Fig. 10.5. For this eigenstate, the boundary condition above is valid because the envelope function satisfies $\phi(x_1) \simeq 0$ even when the infinite potential barriers do not exist. In this way, we can obtain the quasi-bound states of the QW by selecting them out of the full solutions.

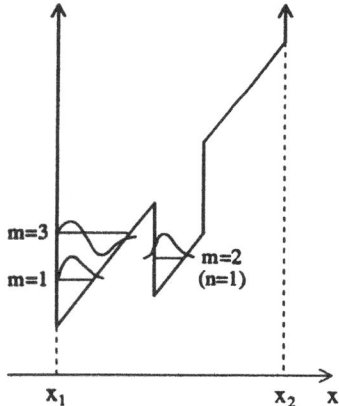

Figure 10.5: Schematic diagram of eigenenergies and envelope functions of a QW subject to an external electric field under the boundary conditions $\phi(x_1) = \phi(x_2) = 0$.

Figure 10.6 shows the field dependence of energy levels in a rectangular QW, where $U_w = 50E_1$, $f = F/F_1$, $F_1 = E_1/eW$ with the electron charge e, $x_1 = -0.8W$, $x_2 = 0.8W$, and the constant effective mass over the whole region is assumed. The results obtained agree well with those reported by Fritz.[14] The envelope functions are also calculated for the same QW with the normalized external field $f = 0, 10, 20, 40,$ and 80, as shown in Fig. 10.7.

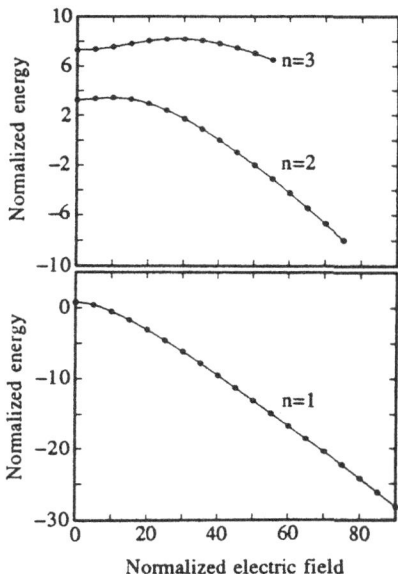

Figure 10.6: Normalized eigenenergy as a function of the normalized electric field for the GaAs/Al$_{0.3}$Ga$_{0.7}$As quantum well.

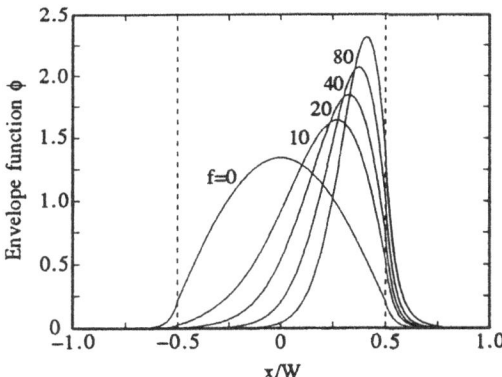

Figure 10.7: Envelope function for the first eigenstate ($n = 1$) of the GaAs/Al$_{0.3}$Ga$_{0.7}$As quantum well.

10.4   Minibands of a Superlattice

First, we consider miniband structures of a rectangular SL in which the well and the barrier have the same thickness, as shown in Fig. 10.8. The calculation was carried out for the GaAs/$Al_{0.1}Ga_{0.9}As$ superlattice with a period of 18 nm.

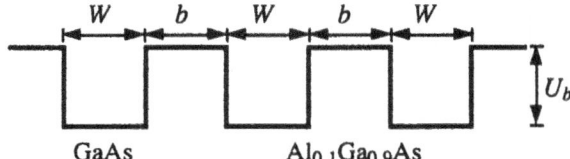

Figure 10.8: Rectangular superlattice.

Figure 10.9 shows the miniband structure of the SL computed by the FEM, together with the exact solutions, where $W = b = 9$ nm and $U_b = 100$ meV. The results are in good agreement. The envelope functions of

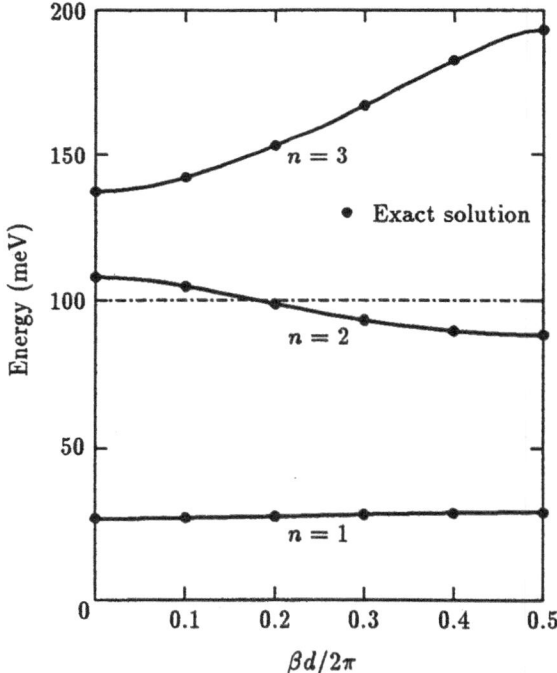

Figure 10.9: Miniband structures for the GaAs/$Al_{0.1}Ga_{0.9}As$ rectangular superlattice.

the electron calculated by the FEM for $\beta = 0$ and $\beta d = \pi$ are shown in

Figs. 10.10(a) and (b), respectively.

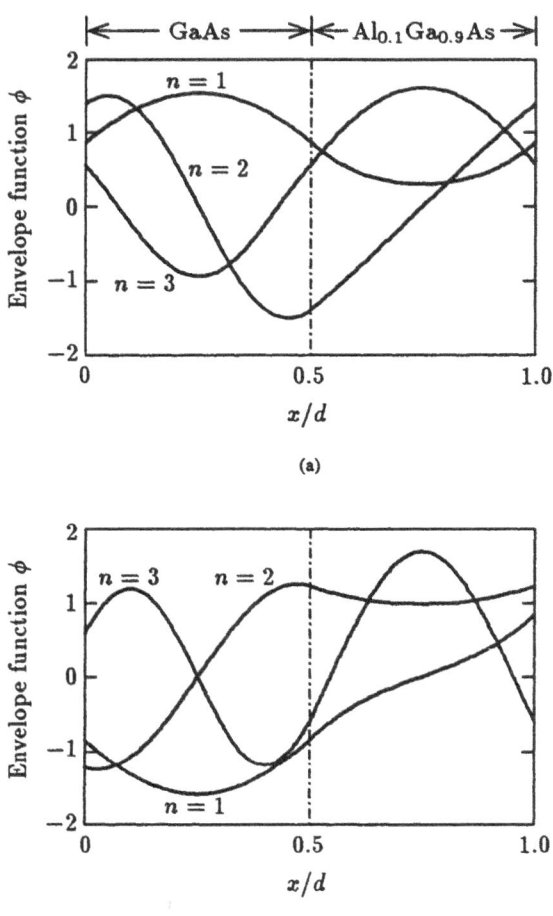

(a)

(b)

Figure 10.10: Envelope functions in the GaAs/Al$_{0.1}$Ga$_{0.9}$As superlattice. (a) $\beta = 0$.
(b) $\beta d = \pi$.

Next, let us consider a bi-periodic structure in which the potential has
two different barriers in a period, as shown in Fig. 10.11. This structure
was proposed by Peeters and Vasilopoulos,[15] for the purpose of controlling
the gap between minibands. The well, the lower barrier, and the higher
barrier are made of GaAs, Al$_{0.3}$Ga$_{0.7}$As, and Al$_{0.4}$Ga$_{0.6}$As, respectively.

Figure 10.12 shows the first two minibands of the structure as a function
of the width of the higher barrier, b_h, where $d = 25$ nm, $b_l = 5$ nm,
$S = 12.5$ nm, $U_l = 228$ meV, and $U_h = 313$ meV. This result is in good

agreement with that reported by Peeters and Vasilopoulos.[15]

In the above calculations, we assumed an abrupt potential change for the heterointerface. In an actual SL, however, an intermediate potential region exists between well and barrier layers. Jiang and Lin[16] reported the calculated results for the miniband structure of an SL with graded interfaces, as shown in Fig. 10.13.

Figure 10.14 shows the miniband structure of the SL for $\Delta = 0$ (abrupt interface) and $\Delta = 0.5$ nm, where $W = b = 6$ nm and $U_b = 322.8$ meV. This result is somewhat different from that reported by Jiang and Lin.[16] This is because we used an effective mass that varies with location (see Eq. (10.4)), while they employed an average effective mass over all region; $m_{\text{eff}} = 0.08 m_0$.

10.5   Resonant Tunneling Phenomena

First, let us consider a simple rectangular potential barrier as shown in Fig. 10.15. Figure 10.16 shows the transmission probability of a GaAs/Al$_{0.21}$Ga$_{0.79}$As rectangular barrier, where $b = 31.6$ nm, $U_b = 158$ meV, where the broken line represents the top of the barrier. The results obtained agree well with the exact solutions.

Next, we consider GaAs/AlGaAs double barrier structures with a rectangular well[17] or a parabolic well[18] as shown in Fig. 10.17(a) or (b), respectively. Figure 10.18 shows the transmission probability of the double barrier structures, where $b = 3$ nm, $S = 10$ nm, $U_b = 956$ meV, where the broken line represents the top of the barrier. These results agree well with those reported by Ando and Itoh,[17] and by Khondker *et al.*[18]

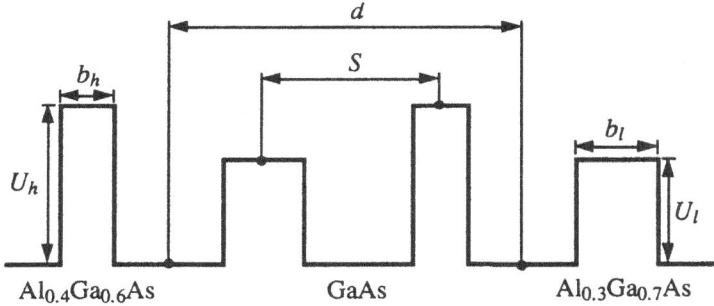

Figure 10.11: Potential profile of a bi-periodic structure.

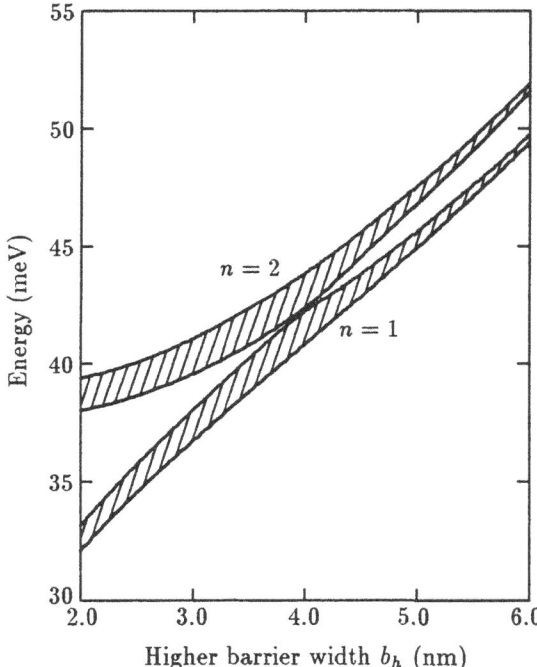

Figure 10.12: Minibands of the bi-periodic structure as a function of the higher barrier width.

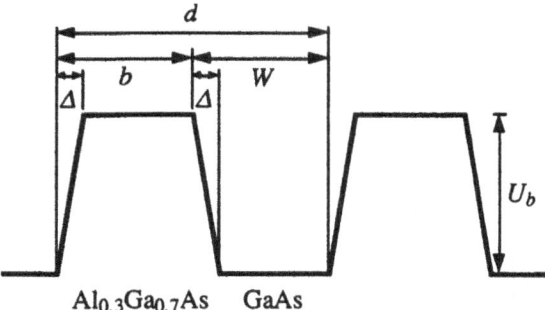

Figure 10.13: Potential profile of a superlattice with graded interfaces.

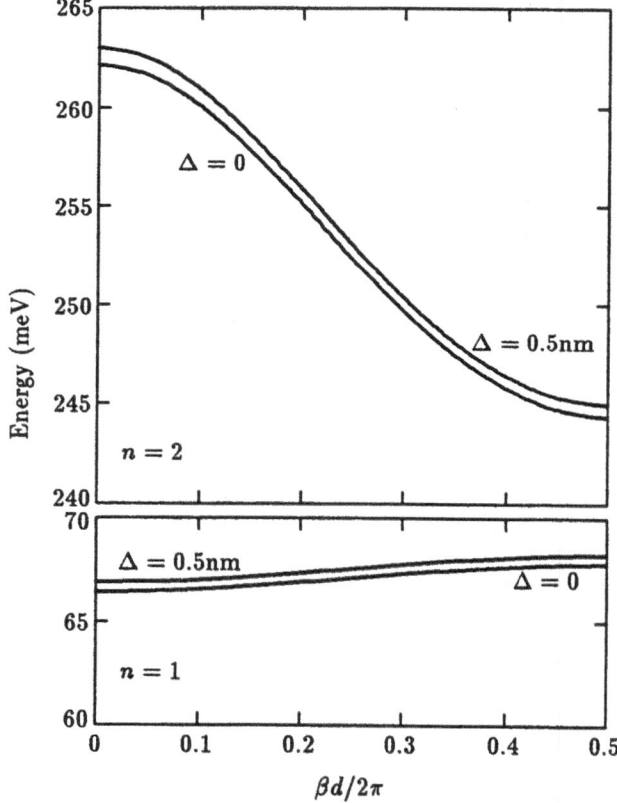

Figure 10.14: Minibands of a superlattice with graded interfaces.

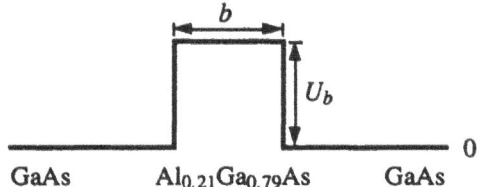

Figure 10.15: GaAs/Al$_{0.21}$Ga$_{0.79}$As rectangular potential barrier.

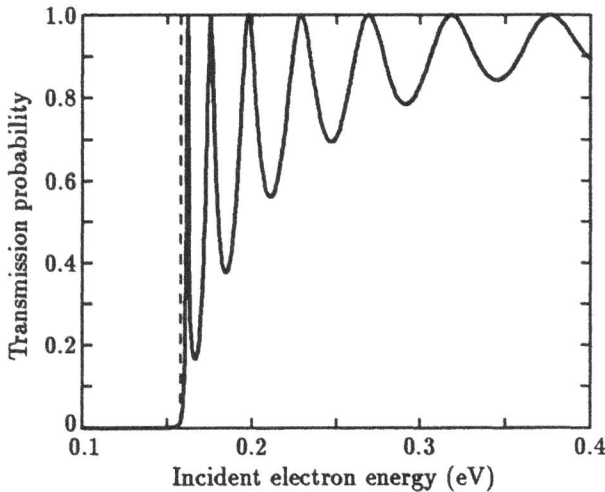

Figure 10.16: Transmission probability for the GaAs/Al$_{0.21}$Ga$_{0.79}$As rectangular potential barrier.

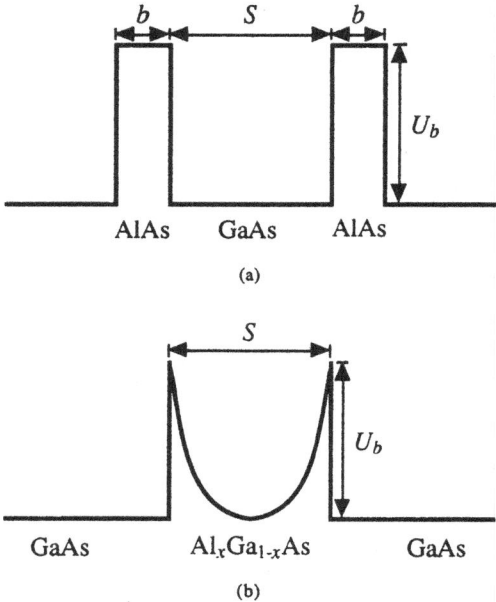

Figure 10.17: Potential profiles of double barrier structures. (a) Rectangular well. (b) Parabolic well.

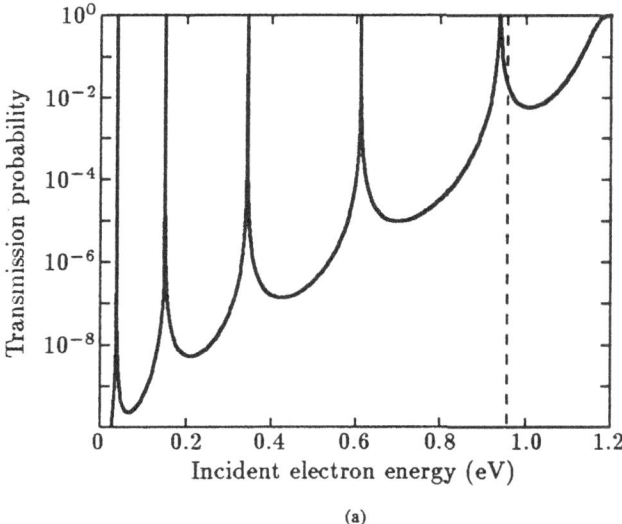

(a)

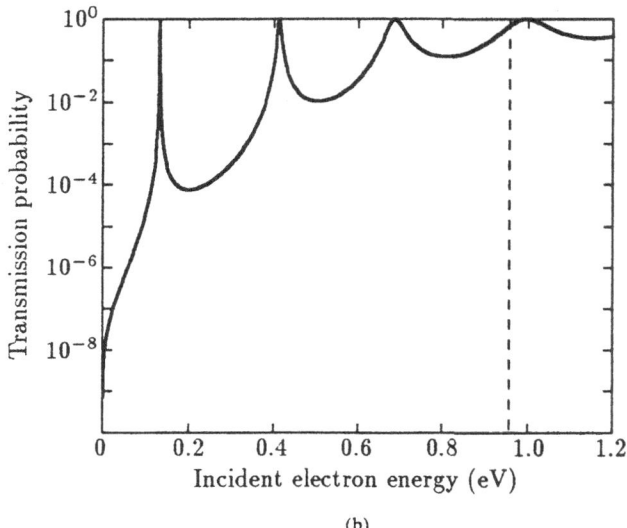

(b)

Figure 10.18: Transmission probability for the double barrier structures. (a) Rectangular well. (b) Parabolic well.

References

1) L. Esaki and R. Tsu, "Superlattice and negative differential conductivity in semi-conductors", *IBM Jour. Res. Develop.*, Vol. 14, pp. 61–65, Jan. 1970.

2) R. Tsu and L. Esaki, "Tunneling in a finite superlattice", *Appl. Phys. Lett.*, Vol. 22, No. 11, pp. 562–564, June 1973.

3) T. Miyoshi, H. Kimura, and M. Ogawa, "Electric field dependence of eigenstates in quantum wells with arbitrary potential distribution", *Trans. Inst. Electron. Inform. Commun. Eng.*, Vol. E70, No. 4, pp. 297–299, April 1987.

4) K. Hayata, M. Koshiba, K. Nakamura, and A. Shimizu, "Eigenstate calculation of quantum well structures using finite elements", *Electron. Lett.*, Vol. 24, No. 10, pp. 614–616, May 1988.

5) K. Nakamura, A. Shimizu, M. Koshiba, and K. Hayata, "Finite-element analysis of quantum wells of arbitrary semiconductors with arbitrary potential profiles", *IEEE Jour. Quantum Electron.*, Vol. 25, No. 5, pp. 889–895, May 1989.

6) K. Kojima, K. Mitsunaga, and K. Kyuma, "Calculation of two-dimensional quantum-confined structures using the finite element method", *Appl. Phys. Lett.*, Vol. 55, No. 9, pp. 882–884, Aug. 1989.

7) K. Kojima, K. Mitsunaga, and K. Kyuma, "Fabrication and characterization of quantum well wires grown on corrugated GaAs substrates by molecular beam epitaxy", *Appl. Phys. Lett.*, Vol. 56, No. 2, pp. 154–156, Jan. 1990.

8) C. S. Lent and D. J. Kirkner, "The quantum transmitting boundary method", *Jour. Appl. Phys.*, Vol. 67, No. 10, pp. 6353–6359, May 1990.

9) C. S. Lent, "Transmission through a bend in an electron waveguide", *Appl. Phys. Lett.*, Vol. 56, No. 25, pp. 2554–2556, June 1990.

10) L. R. Ram-Mohan and J. Shertzer, "Electronic energy bands and optical nonlinearity of checker-board superlattices", *Appl. Phys. Lett.*, Vol. 57, No. 3, pp. 282–284, July 1990.

11) C. S. Lent, "Ballistic current vortex excitations in electron waveguide structures", *Appl. Phys. Lett.*, Vol. 57, No. 16, pp. 1678–1680, Oct. 1990.

12) K. Nakamura, A. Shimizu, M. Koshiba, and K. Hayata, "Finite-element calculation of the transmission probability and the resonant-tunneling lifetime through arbitrary potential barriers", *IEEE Jour. Quantum Electron.*, Vol. 27, No. 5, pp. 1189–1198, May 1991.

13) K. Nakamura, A. Shimizu, M. Koshiba, and K. Hayata, "Finite-element analysis of the miniband structures of semiconductor superlattices with arbitrary periodic potential profiles", *IEEE Jour. Quantum Electron.*, Vol. 27, No. 8, pp. 2035–2041, Aug. 1991.

14) I. J. Fritz, "Energy levels of finite-depth quantum wells in an electric field", *Jour. Appl. Phys.*, Vol. 61, No. 6, pp. 2273–2276, May 1987.

15) F. M. Peeters and P. Vasilopoulos, "New method of controlling the gaps between the minibands of a superlattice", *Appl. Phys. Lett.*, Vol. 55, No. 11, pp. 1106–1108, Sept. 1989.

16) H. X. Jiang and J. Y. Lin, "Band structure of superlattice with graded interfaces", *Jour. Appl. Phys.*, Vol. 61, No. 2, pp. 624–628, Jan. 1987.

17) Y. Ando and T. Itoh, "Calculation of transmission tunneling current across arbitrary potential barriers", *Jour. Appl. Phys.*, Vol. 61, No. 4, pp. 1497–1502, Feb. 1987.

18) A. N. Khondker, M. R. Khan, and A. F. M. Anwar, "Transmission line analogy of resonance tunneling phenomena: The generalized impedance concept", *Jour. Appl. Phys.*, Vol. 63, No. 10, pp. 5191–5193, May 1988.

INDEX

267

The manufacturer's authorised representative in the EU is Springer
Nature Customer Service Centre GmbH, Europaplatz 3, 69115 Heidelberg,
Germany. If you have any concerns regarding our products, please
contact ProductSafety@springernature.com

Printed and bound by CPI Group (UK) Ltd, Croydon, CR0 4YY
23/04/2026
02095625-0003